S0-CAI-930

Integrated Cost Management

Integrated Cost Management

A Companywide Prescription for Higher Profits and Lower Costs

Michiharu Sakurai
Senshu University
Tokyo, Japan

Productivity Press
Portland, Oregon

Written in collaboration with D. Paul Scarbrough, Brock University, Ontario, Canada

Productivity Press
P.O. Box 13390
Portland, OR 97213-0390
United States of America
Telephone: 503-235-0600
Telefax: 503-235-0909
E-Mail: service@ppress.com

Cover design by William Stanton
Composition by Caroline Berg Kutil
Graphics by Productivity Press (India) Private Ltd.
Printed and bound by Edwards Brothers in the United States of America

Library of Congress Cataloging-in-Publication Data

Sakurai, Michiharu.
Integrated cost management: a companywide prescription for higher profits and lower costs / Michiharu Sakurai; in collaboration with D. Paul Scarbrough.
p. cm.
Includes bibliographical references and index.
ISBN 1-56327-054-4
1. Cost accounting. 2. Activity-based accounting. 3. Reengineering (Management)—Japan. 4. Reengineering (Management)—United States. I. Scarbrough, D. Paul. II. Title.
HF5686.C8S238 1996
658. 15'52—dc20 95-41669
CIP

00 99 98 97 96 95 10 9 8 7 6 5 4 3 2 1

Contents

Publisher's Message

In the past, accounting has served primarily as a tool for recording company history and has attempted through budgets and standards even to project a company's financial future. Today, however, through the understanding and efforts of people like Professor Michiharu Sakurai, accounting is becoming a valuable partner in managing the present.

As professor at one of Japan's top business universities and one of two board members on Japan's CPA evaluation committee, Dr. Sakurai has been responsible for updating the Japanese CPA exam and continues to be a leader in articulating the latest thinking in both taxation and management accounting. A year ago in a survey led by Japan's leading accounting journal, 300 accounting professors were asked to rank their peers, and Dr. Sakurai was rated number one in the country. He is one of the few people to bridge Japanese and American accounting and management systems. Both through his depth of knowledge and through his relationships with leaders in both societies, Dr. Sakurai has become a primary source of understanding about the new role that accounting must play in managing a company in a global economy.

In his latest work, *Integrated Cost Management,* Dr. Sakurai addresses the need for integration of all processes and describes a strategy for that integration as the required next step in setting goals for world class organizations. The book begins with a review of Japanese management accounting practices and changes that have led to the need for integration, followed by a discussion of both Japanese and western approaches to measurement and the impact of factory automation and computer integrated manufacturing on business management. He clarifies the primary issues related to major cost management approaches, discusses the value of target costing in planning product design and production processes, and includes a discussion of performance evaluation for high technology companies. He concludes by addressing the future of management accounting as a primary means to effect and enhance a company's global presence. The primary emphasis throughout the book is an articulation of the leadership role of the management accountant in formulating strategy for a company's integration and globalization.

It has been my great pleasure to know Michiharu Sakurai for many years and to publish his fine work in English. As a friend and admirer, I am grateful to publish this latest book, which refines his thinking of the past and points to the future of management accounting. I also want to thank the staff of fine people who prepared the book for publication: Diane Asay, editor in chief; Susan Swanson for prepress coordination; Bill Stanton for cover and text design coordination; Aurelia Navarro for proofreading; Caroline Berg Kutil for typesetting; and Edwards Brothers for printing and binding.

Norman Bodek
Publisher

Foreword

Recent developments in management accounting have increased the overlaps among disciplines. For example, periodicals and books on management accounting cover such topics as quality, innovation, and speed. Some articles could just as easily appear in general management or engineering publications instead of accounting publications. I applaud these developments. The more interdisciplinary accountants become, the better equipped they will be for coping with incessant change. Most important, to survive and thrive, accountants must examine their roles from cross-functional, general-manager, and user viewpoints.

I met Professor Michiharu Sakurai over twenty years ago when he came to Stanford University for a short visit. Along with Professor Kiyoshi Okamoto, he was a gracious host during my 1991 visit to Japan. Our friendship grew in 1992 and 1993 when he made an extended visit to Stanford.

Professor Sakurai has a broad and deep understanding of management accounting theory and its practical applications in Japan and the United States. To my knowledge, he understands the major differences in practices between our two countries as well as or better than any other researcher.

The topics in this volume cover the broad spectrum of management accounting. Professor Sakurai draws on his extensive knowledge of both U.S. and Japanese companies and literature. He provides informative comparisons of development in each country regarding the variety of management accounting topics. For example, in Chapter 5's discussion of activity-based management (ABM) and activity-based costing (ABC), he contrasts the tendencies of Americans to focus on measurement and of the Japanese to focus on management. He makes similar observations regarding the management of quality and its costs.

Professor Sakurai provides especially thorough discussions of such central management accounting topics as target costing, overhead management, and investment justification in computer-integrated management (CIM). In addition, he offers a penetrating study of the cost management of software, an increasingly prominent cost in our global economy.

Furthermore, serious readers will welcome the ample bibliographies accompanying each chapter. These include references to articles by authors from many nations including Japan.

Integrated Cost Management enriches the literature of management accounting by providing a sweeping perspective on the strengths and weaknesses of Japanese concepts and techniques. In addition, Professor Sakurai provides ample comparisons and reasons for divergencies between American and Japanese development and practices. In this way, management accountants throughout the world will learn how managers can run their organizations better than ever.

Charles T. Horngren
Edmund W. Littlefield Professor of Accounting
Graduate School of Business
Stanford University

Preface

The purpose of this book is to help readers develop new management accounting systems suited for today's advanced manufacturing technology. The systems now used by leading-edge companies in the United States and Japan provide the starting point for discussion. However, my ultimate purpose is not simply to make suggestions and comments on existing practices, but to help advance management accounting through a more critical analysis of our practices. To do so, we need to recognize the differences and similarities in management accounting between our two nations.

Japanese Management Accounting

Forty years ago management accounting in the United States and Japan was essentially the same. At that time Japanese companies were actively introducing new management accounting concepts and techniques into Japan. While these new methods were primarily of U.S. origin, there was also a substantial German influence. However, by the late 1960s Japanese businesspeople had grown dissatisfied with the imported methods and were designing their own style of management accounting systems.

Japanese markets, business practices, historical development, and culture were quite different from those in the United States. This meant that U.S. methods were not always appropriate in Japan. These dissimilarities had become more pronounced by the late 1960s as higher standards of living in Japan were accompanied by an increasing diversity in consumer desires. Japanese consumers sought quality products with greater individuality.

As we will discuss in Chapter 1 and elsewhere, the response of Japanese businesses was to produce a relatively low volume of more diversified products (high variety/low volume production). This tactic could be viable only with the rapid development of flexible manufacturing systems (FMSs). And, in fact, factory automation (FA) is now the main production regime in better companies regardless of size. In the 1970s and 1980s, quality improvement programs and just-in-time (JIT) activities along with widespread use of FA brought Japan rapid productivity increases and economic development.

Production is most effective as a tool for attaining corporate goals when it can respond flexibly and quickly to customer needs. The use of point-of-sale (POS) systems to record interactions with customers and inject the resulting knowledge into internal communication networks allows managers to effectively link consumer needs to new product development. This linkage of consumer, designer, and producer—computer-integrated manufacturing (CIM)—is sweeping through Japan with unprecedented speed. Furthermore, significant alteration in management accounting techniques accompanied all of the changes listed here.

Understanding Other People's Methods

Cumulatively, these factors mean that consumer and producer markets and production regimes in the United States and Japan are very different today. This has led to substantial differences between their respective systems of management accounting. However, the full extent of disparity is not always perceived by observers largely because our shared heritage of earlier methods tends to bury changes in practice under similarities of language and form. This is especially common when Japanese and American accountants interact because the Japanese are more often the bilingual participants. Being intimately familiar with U.S. practices, they often translate their own ideas into the nearest standard U.S. accounting equivalent rather than give an extended description of their actual methods. The monolingual American accountants often fail to realize that they are hearing the *Classics Illustrated* version of Japanese practice. This type of imprecise, one-way translation conceals as much as it reveals. Although this phenomenon cannot be totally avoided even in this book, great pains have been taken to pierce the surface of the accounting methods and reveal what lies beneath.

The Japanese Economy Today

As I write, Japanese markets are in turmoil as the nation struggles to recover from the deep recession caused by the rapid increase in the value of the yen, the collapse of the bubble economy, and low consumer demand due to poor purchasing power. Of these three major forces, the too-rapid and wide-impacting increase in the value of the yen is thought by most economists to be the dominant force. One dollar bought 360 yen before the 1960s, 260 yen in 1975, 127 yen in 1988, and 80 yen in April 1995.

A change of such magnitude and speed can have striking and unpredictable effects on any nation, but especially on a resource-poor but internationally involved nation like Japan. Japan must buy and sell virtually everything in a bazaar of shifting foreign currencies. This has led to remarkable but erratic changes in the cost structure of individual companies as well as in the nation as a whole. Economists admit that it is high time to restructure the organizations and reengineer the business processes of Japanese companies. While certain high-profile companies excel, on average Japanese companies are not as efficient as their counterparts in the United States.

In the late 1980s, Americans created TQM after studying TQC in Japanese companies. Today Japanese businesspeople are taking a new and serious look at the U.S. management accounting systems recently developed, particularly activity-based costing (ABC) and activity-based management (ABM). This mutual inspection is the beginning of what I hope is a fruitful interaction. I would be very happy if this book contributes to this process.

Michiharu Sakurai

Acknowledgments

Several prominent American professors and universities were instrumental in giving me the intellectual background needed for this book. I owe a great deal to three professors and their colleges. First, Professor Larry N. Killough allowed me to visit the Virginia Polytechnic Institute and State University (VPI) from April 1983 to March 1984. My wife Hiroko and my son Toshi join me in sincerely thanking the VPI faculty and staff for receiving us so warmly and for introducing us to many good friends—Professors Wayne E. Leininger, Konrad M. Kubin, James O. Hicks, and Robert M. Brown to name a few.

Second, Professor Robert S. Kaplan brought me to Harvard Business School as Senior Fulbright Visiting Scholar from September 1989 to March 1990. For his thoughtfulness in giving me the opportunity to discuss activity-based costing with him and with Professor Robin Cooper, I am deeply grateful.

Third, Professor Charles T. Horngren afforded me the opportunity to visit Stanford University from November 1992 to February 1993 where I had pleasant and informative discussions of management accounting with Professors George J. Foster and Srikant M. Datar. In fact, I would have been unable to write this book had I not met these excellent teachers.

Most importantly, I wish to acknowledge my good friend Professor D. Paul Scarbrough of Brock University (St. Catharines, Ontario), who reviewed the manuscript many times and provided valuable input. Without his assistance, I would have been unable to complete this book. Discussions on financial organization and company visits with Professor Patrick J. Keating of San Jose State University and Stephen F. Jablonsky of the Pennsylvania State University gave me a good opportunity to have an insight into management accounting practices in the U.S. and Japan. I also gratefully recall my visits to a dozen American and Japanese companies between October 1991 and April 1992. Likewise, the following professors receive my special thanks for their worthy advice: Professor Eric Noreen, University of Washington (Chapter 8); Professor Philip Huang, VPI (Chapter 6); lecturer John Fahy, Trinity College (Chapter 12); and lecturer David A. Trokeloshvili, Nippon University (English editing). And I again acknowledge Bob Kaplan for his excellent advice on the final version of Chapter 5 (ABC/ABM).

My research and conclusions are based on the successes and experiences now being realized by a handful of Japanese organizations. I owe a great deal to these companies and their far-sighted executives and managers. As always, I thank my colleagues at Senshu University for their support and advice. In addition, Senshu University granted me financial support in the form of a subsidy for the publication of this book.

Lastly, Productivity Press must be acknowledged for giving me the opportunity to write this book for you. I am especially grateful to Mr. Norman Bodek, publisher; Ms. Cheryl Rosen, senior editor; and Ms. Karen Jones, managing editor, for their professionalism and patience.

Integrated Cost Management

CHAPTER 1

Current Management Accounting Practices in Japan

This chapter presents a framework for understanding how management accounting is currently practiced in Japan in relation to past developments in both Japan and the United States. The framework, evolved from the 1993 Special Committee Report of the Japan Accounting Association, will be used throughout the book as a basis for discussion and analysis. It represents the management accounting practices of the typical company in Japan.

During the 1950s and early 1960s, the United States dominated the world economy. In those years, Japanese management accountants learned a great deal from their American counterparts. Some of the methods studied were: responsibility accounting, capital budgeting, accounting for capacity cost, direct standard costing, contribution margin analysis, accounting for decision making, linear programming and profit planning, zero-base budgeting, behavioral accounting, and social responsibility accounting. Most of these management accounting tools evolved in the United States after World War II. Most were actively studied and many were implemented in Japanese companies (much as Japanese methods were studied by the Americans in the late 1980s). At the time, Japanese businesspeople believed that American companies held the key to effective production methods.

There has been less to learn from the United States since the 1970s. Johnson and Kaplan note this scarcity of relevant research in their book *Relevance Lost* (1987). It may be the result of too great a focus on capital markets research by U.S. accounting academics seeking degrees, tenure, or promotion instead of a contribution to society. However, Kaplan's works have motivated U.S. accountants to experiment with management accounting tools such as activity-based costing (ABC). Today American managers increasingly focus on advanced cost management methods such as total quality management (TQM) and target costing, as well as ABC.

Since the 1980s when Japan began to lead the world in factory automation (FA), the manufacturing environment has changed rapidly. For example, robotics was introduced aggressively during the last decade. As a result, of the 1993 world robot population of 610,000, 60 percent (368,000) operated in Japan with only 8 percent (50,000) in the United States (Japan Robot Association, 1994).

The shift from flexible manufacturing systems (FMSs) to factory automation (FA) and now to computer-integrated manufacturing (CIM) has necessitated adjustments to management accounting practices. In the 1990s, with the development of unique management accounting and engineering tools like target costing and just-in-time (JIT), Japan's management accounting framework has broken away from that of the United States.

Changes in Corporate Goals and Management Accounting Techniques

Today, Japanese companies need effective management. That is, to reestablish their economic strength, Japanese companies must change their emphasis from *volume expansion* to *effective management*. The following is a brief review of the history of Japanese economic development and the major management accounting techniques used in each period.

From Postwar to 1950s

The major goal or mission of the day was improving efficiency as well as improving quality. Japanese managers tried to increase input resource efficiency by using standard costing and other management accounting tools. Standard costing was the most effective tool for improving the efficiency of manufacturing production during those days. Consequently, major Japanese companies very actively implemented standard cost systems during this period.

From 1960s to 1991

The major mission of this period in typical Japanese companies was volume expansion. This era can be divided into two periods: from the 1960s to 1973 (the first oil crisis) and from 1973 to 1980s. In 1991, the bubble economy of the late 1980s collapsed.

The period from the 1960s to 1973 is characterized by high economic growth. Japanese firms developed new markets through mass-production and resultant cost reduction. Economy of scale was gained by volume production. Variable costing was the most popular tool during those days. It was introduced primarily into process-oriented industries throughout Japan. Variable costing was popular because it provided managers with tools for determining what to produce when there was idle capacity. Some Japanese companies sold their products even if they could not expect enough return. Under these conditions, return on investment (ROI) could not be a major goal for typical Japanese companies.

The period from 1973 to the 1980s is characterized by stable or low economic growth. The oil crises taught some Japanese managers that volume expansion alone may not be appropriate for the future. A few excellent companies realized that effective use of resources must be pursued instead of volume production. For example, quite a few Japanese assembly-oriented companies introduced target costing. Instead of mass-production, these companies turned to low volume production with varied products. In place of scale, economy of scope played an important role. In such situations, target costing is an appropriate tool for the effective use of materials and parts in assembly-oriented industries. However, the majority of Japanese companies did not realize that volume expansion was no longer appropriate as a major goal. On the contrary, the typical Japanese company still pursued a business strategy of volume expansion during this period, as evidenced by the active investment in domestic plants and equipment and overseas operations of the late 1980s.

From 1991 to Today

The Japanese economy was staggered by the collapse of the bubble economy. The ensuing high yen has devastated exporting companies. Even successful companies could not export domestically produced products competitively, and were forced to produce products overseas. Heavy overhead is challenging formerly competitive industries. Factory overhead along with general and administrative expenses increased dramatically during the bubble economy in the belief that growing sales would make them recoverable. As a result, the break-even point has risen alarmingly in typical Japanese manufacturing firms.

There will be little or no possibility for Japan to enjoy high economic growth in the future. Many economists predict that Japan may suffer greatly from the large, and growing, pension obligations created as Japanese society ages. The increase in external surplus and the resultant higher value of the yen increases friction with other industrialized countries, and drives Japanese companies towards direct overseas investment. Thus, it accelerates industry-hollowing in Japan. The gradual decrease in the number of young workers and deterioration of their motivation (*kinrouiyoku*) may foreshadow a deterioration of worker's quality.

The *keiretsu* (affiliate companies, or large scale business combines) must be decreased or destroyed because they are targeted as an unfair business practice by other countries. This threatens the good relations with suppliers that has characterized Japan's economic success.

Let's summarize the change of corporate goals by using the input/output assumptive example shown in Figure 1-1. During the efficiency improvement period (from postwar to 1950s), Japanese tried to reduce inputs (for example, from 120 to 100) in order to get an output of 150. During the volume expansion period (from 1960 to 1991), the Japanese increased input to 200 to get 270. As a result, return decreased to 135 percent from 150 percent. Until the oil crisis of 1973, Japanese companies expanded market share and sales volume by mass production. Since the oil crisis they have changed their policy from mass production to low volume production with varied products, but their mindset was still the same—volume expansion. This approach may, in fact, have caused the bubble economy.

Year	Postwar – 1950s	1960 – 1991	1991 – Present
Major goals	Efficiency improvement	Volume expansion	Effective management
$\frac{\text{Output}}{\text{Input}}$	$\frac{\text{Output}\uparrow}{\text{Input}\downarrow}\left(\frac{150}{120 \rightarrow 100}\right)$	$\frac{\text{Output}\uparrow}{\text{Input}\uparrow}\left(\frac{270}{200}\right)$	$\frac{\text{Output}\uparrow + \alpha}{\text{Input} \rightarrow}\left(\frac{300 + \alpha}{> 200}\right)$

Figure 1-1. Changes in Japanese Corporate Goals

"Effective" versus "Efficient" Management

With effective management as a goal, Japanese must get 300 units of output or more by using 200 units, or less, as input. If we examine only the input/output relationship, we can use the term "efficiency." However, the contents of the output must be changed in the future, requiring a shift to "effectiveness." In Figure 1-1, output under "effective management" includes "+ α." This concept includes social benefits that cannot be quantified, such as:

- reduction of working hours
- environmental preservation
- good relationships with suppliers
- customer satisfaction
- protection of shareholders

These + α items can be expressed as goals but they can also be treated as constraints or costs to individual companies. In order to increase efficiency, Japanese management will be required to get economies of speed by reducing manufacturing cycle time, delivery, turnover, and throughput.

In the present state of the Japanese economy, Japanese managers suffer from the high yen and the decline in price of almost every commodity except labor. Therefore, it may be impossible for them to get 300 units of revenue. They must reduce prices to maintain the present sales volume, so that they can get only 260 or 240 revenue with 300 output. Then, they must reduce input to 160 or 140 to improve the return. To reduce input resources Japanese managers must operate their businesses in more efficient ways. As a result of these constraints they must focus on pursuing effective management. And to pursue effective management, the Japanese need integrated cost management.

Cost Management versus *Integrated Cost Management*

Throughout this era of change, two documents have provided the basis for the evolution of Japanese management accounting. These are *Cost Management* (1967) and *Integrated Cost Management* (1993). We shall examine both.

Cost Management

In 1967 Japan's Ministry of International Trade and Industry (MITI) published *Cost Management.* This report represented the management accounting philosophy and techniques of the day. In addition, it played a leading role in the further development of Japanese management accounting theory and practice. Previous to this report, the 1962 Cost Accounting Standards (Ministry of Finance, 1962) guided Japan's cost accountants. These standards integrated standard costing with the accounting system to control costs and prepare financial statements.

Cost Management had three unique characteristics for the time. First, it stressed the role of cost planning as well as cost control. In other words, it clearly indicated the limitations of standard costing as a tool for cost management and suggested the need for new tools for cost planning. Second, it emphasized increasing revenue as well as reducing costs. Third, it advocated establishing cross-functional responsibility accounting in major Japanese companies. The effect of these three factors on the direction of Japanese management accounting in the following years was significant.

Integrated Cost Management

In 1993 the Japan Accounting Association established a special committee to study cost management systems for the new business environment. When their findings were published in a report entitled *Integrated Cost Management* (1993), they represented the most influential philosophies and techniques of Japanese management accounting in practice.

"Integrated cost management" (ICM) is defined as a comprehensive value-chain approach to strategic cost management for products, software, and services. It covers the entire product life cycle from research and development (R&D) to product planning, design, manufacturing, sales promotion, physical distribution, operation, maintenance, and disposal. Cost reductions/quality improvements across the life cycle are expected to provide the highest organizational benefit. ICM emphasizes the preeminence of attaining corporate goals in the globalized competitive world.

The framework described in the 1993 report provides the basis for this chapter. For comparison, Table 1-1 contrasts the typical philosophies and techniques of the 1960s with those of the 1990s.

Table 1-1. Management Accounting in the 1960s and 1990s

Description	1960s	1990s
Business environment	Export promotion Mass production (Low-variety, high-volume) Industrialization	Globalization Flexible production (High-variety, low-volume) Computerization
Corporate mission	Profit	Survival, growth, and development
Corporate goals	Profitability	Effectiveness
General business approach	Planning & control	Planning and control, plus innovation, kaizen, and maintenance
Organizational structure	Functional	Cross-functional
Major areas that use management accounting	Production and marketing	R&D, planning and design, marketing, operations, maintenance, and disposal
	Production	Production, service, software
Primary management accounting techniques used	Standard costing Budgeting Variable costing Operations research (OR) Industrial engineering (IE) Others	Traditional techniques of the 1960s, plus Target costing ABC & ABM Quality costing Life-cycle management TQC, TPM, JIT, VE Others

Areas Addressed by Integrated Cost Management

Japanese business is embedded in a social structure that has changed greatly since the 1960s. These changes extend to the business environment, business philosophy, and corporate goals.

Business Environment

Three characteristics differentiate the Japanese business environment over the past thirty years—globalization, factory automation, and computerization or information technology. Each has had great impact.

Globalization

Driven by its scarcity of natural resources, Japan's economy developed around the MITI-promoted philosophy of producing and exporting quality products at low cost. In the 1960s, one U.S. dollar was the equivalent of 360 yen. However, since the 1985 Plaza Agreement, the yen's value rose from 260 yen per dollar (February 1985) to 127 yen (January 1988) and then to 80 yen (April 1995). Japanese reaction was to switch policies: "Produce in Japan, Export Abroad" rapidly became "Produce Abroad, Sell Abroad." As a result, numerous Japanese companies built overseas plants and stopped manufacturing products at home.

This, of course, parallels what many business theorists suggest is the life cycle of international business, an issue that we will examine in Chapter 12.

Factory automation

Over the last several decades, increases in the average household income greatly affected Japanese lifestyle. Diversifying consumer demands and shortened product life cycles altered the economic pattern to such an extent that production methods were overhauled. The standard switched from mass production and limited product variety to small lot production and enormous product variety. Indeed, consumer demand was one of the strongest forces behind the development of flexible manufacturing with industrial robots. Major Japanese companies continue to move in this direction by introducing CIM in their factories.

Computerization

Computerization coupled with new communication and information technologies has hastened a change worldwide from "industrial culture" to "information culture." This has prompted business innovations in Japanese organizations (and likewise in the United States), developments pushed along by the increasing need for global communication in newly globalized companies.

Corporate Mission

The major corporate mission or business philosophy of the 1960s was to make a profit by expanding business operations (volume expansion)—although Japanese managers were known to use multiple goals. For example, goals of increasing sales volume, market share, and productivity were coupled with maintaining good relationships with employees, banks, investors, and suppliers. Gradually, the primary mission shifted to include greater social dimensions. The significance of this shift is Japanese management's realization that business must be done in harmony with society.

Currently, profit or other volume-oriented corporate goals alone do not dominate. Many Japanese managers believe that the highest corporate philosophy or mission is not to make a profit; instead, it is the survival, growth, and development of the organization.

Corporate Goals

The 1960s brought huge demands to satisfy in both domestic and foreign markets. Major corporate goals were to raise sales volumes by increasing market share at home and abroad. The

1967 *Cost Management* report pointed in this direction. However, partly due to saturated consumer demands, the market is no longer growing. As a result, most Japanese companies cannot reasonably expect to achieve higher profitability by efficiencies that increase sales or market share.

Instead, the current keyword is effectiveness—the effective use corporate resources for the survival, growth, and development of the organization. Effectiveness means the wise use of resources to achieve quality, flexibility, service, delivery, throughput and so on—along with sales growth and market share. This relationship is shown in Figure 1-2.

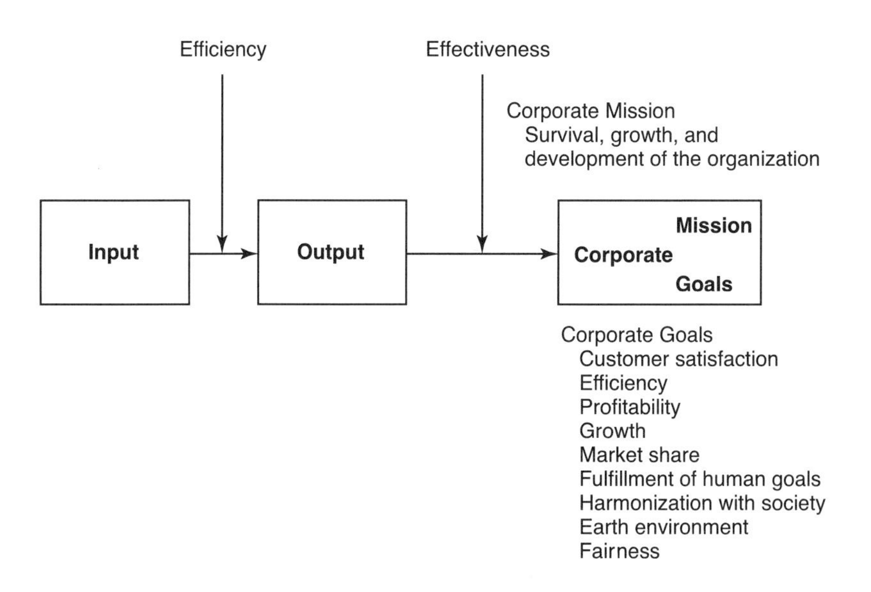

Figure 1-2. Changes in Corporate Mission and Corporate Goals

To achieve the goals of survival, growth, and development, business goals must change. Customer satisfaction, efficiency, profitability, growth, market share, fulfilling human goals, harmony with society and the physical environment, and fairness are recognizable current corporate goals.

Responding to these factors, management accounting practices in Japan have gradually switched from predominantly U.S.-derived methods to distinctly Japanese approaches. This change has penetrated the general business approach, organizational structure, major areas of accounting activities, and accounting techniques.

General Business Approach

The earlier *Cost Management* report uniquely focused on planning as well as cost control. It emphasized the limits of standard costing as a tool for management control by acknowledging its retrospective nature. The more recent *Integrated Cost Management* goes beyond planning and control to emphasize "innovation, kaizen, and maintenance" (IKM). In this context, innovation means innovative changes to products or production processes as a result of introducing new technology and/or investing in plant and equipment. It is the result of discontinuous discovery activities and is radical in its effect. Kaizen is continuous improvement of current activities. Maintenance refers to activities for maintaining current standards in technology, business, and operations. (We will examine the IKM framework further in Chapter 3.)

The IKM approach is explicitly purposeful—from the very names of the three areas to the language used. On the other hand, the planning and control approach more characteristic of American companies is not. The planning and control approach assumes (1) that an organization's purpose comes from top management and (2) that the tools and their users are somewhat detached professional service providers. None of the American tools listed at the beginning of this chapter implies improvement in any business function. Instead, they imply management planning and control. In fact, it would be possible to have a perfectly functioning system using those techniques without having any improvements or even attempts at improvements.

The IKM framework popular in Japan is believed to be a more suitable operating doctrine than the planning and control approach for companies using total quality control (TQC), just-in-time, total productive maintenance (TPM), and target costing. They do not seek to control the subject—but to change its nature. TQC's purpose is to improve quality. JIT's purpose is to reduce inventory. TPM's purpose is to improve machine availability. Target costing's main purpose is strategic cost management. The IKM operating doctrine and these Japanese tools probably developed together. However, even if they did not, they work best together.

It is not widely understood by Americans that this attitude of willingness to undertake fundamental change is the driving force behind Japanese management tools like the ones just mentioned. In fact, this change in approach may be the most significant piece of the puzzle.

Organizational Structure

Traditionally, business organization revolved around functions such as production, marketing, purchasing, R&D, engineering, personnel, and accounting. This functional organization is still essential because of the knowledge-base developed by each area. However, cross-functional structures are also indispensable for developing new products or conducting research and development. They allow the cooperative work of one area to influence another area—besides allowing tradeoffs in cost. Even though the information density in each functional area has increased dramatically, this influence often occurs informally. At the same time, informal communication has become more difficult as physical and organizational distances increase in larger firms. To ensure that cross-functional activity takes place even in the face of obstacles, many companies create formal cross-functional work groups.

Major Areas That Use Management Accounting

Major areas that use management accounting have changed along with the other changes discussed in this chapter. In the past, management accounting was used primarily in production, marketing, and sometimes finance functions. Currently, it is used in many other areas; for instance, in R&D, product planning and design, and life-cycle planning areas like operations, maintenance, and disposal. Whereas in the United States we would expect this to cause an increase in the number of accountants, this has not been the case in Japan due to certain features of its business environment.

In the past, management accounting's usefulness was confined to the manufacturing industry. Today it is used extensively in computer software, telecommunications, merchandising, and other service industries.

Primary Management Accounting Techniques

In the 1960s and early 1970s, traditional management accounting techniques like standard costing, operating budget, variable costing, and capital expenditure budget were employed in much the same way by Japanese and American companies. This was a result of the introduction of U.S. management accounting systems into Japan following World War II. Although well entrenched in Japanese companies, these traditional techniques often underwent modification to fit Japanese preferences.

Today, a set of new tools appears on the managerial accounting palette. So far, new tools such as target costing, kaizen costing, and cost maintenance have been used most intensively in assembly-oriented industries—although their popularity spreads rapidly. Cost engineering tools such as TQC, JIT, TPM, and value engineering (VE) have also been installed effectively in these companies. These latter tools are regarded as distinct from—but compatible with—managerial accounting tools. Their use extends even to Japanese software houses and mainframe manufacturers, who have installed cost accounting and cost management systems for software that include both managerial accounting and cost engineering tools.

While actively introducing techniques like quality costing, life-cycle management, chargeback systems, and investment justification for CIM equipment, Japanese companies have also developed management accounting systems for globalized companies. In late 1993, with Japan's recession deepening, they even began to show great interest in activity-based costing (ABC) and activity-based management (ABM) as tools for business process reengineering.

Conclusion

This discussion of the Japanese business environment provides a basis for our examination of the management accounting concepts and techniques in leading-edge companies that appear in the remainder of this book. The overriding imperatives of competition in the globalized business world are responsiveness and flexibility as Johnson indicates in *Relevance Regained* (1992). Integrated cost management is expected to enhance bottom-up empowerment, enabling an organization to be responsive and flexible in the globalized world.

Since one of the most influential forces behind the new concepts and techniques for Japanese management accounting is factory automation, we will discuss the impact of FA and CIM on Japanese business management in Chapter 2.

References and Further Readings

Japan Robot Association. 1994. "The Present State and Outlook for Industrial Robots: October 1994." Japan Robot Association, p. 16.

Johnson, H. Thomas. 1992. *Relevance Regained: From Top-Down Control to Bottom-Up Empowerment.* New York: The Free Press, p. 16.

Johnson, H. Thomas, and Robert S. Kaplan. 1987. *Relevance Lost: The Rise and Fall of Management Accounting.* Boston: Harvard Business School Press, pp.1,12.

Ministry of Finance (Business Accounting Council). 1962. Cost Accounting Standards.

MITI, Deliberation Council on Industrial Structure. 1967. *Cost Management.* MITI.

Sakurai, Michiharu. 1995. "Past and Future of Japanese Management Accounting." *Journal of Cost Management* (fall), pp. 19-28.

Special Committee Report on Constructing a Cost Management System for the New Business Environment. 1993. "Integrated Cost Management." *Proceedings of the 1993 Japan Accounting Association* (September), pp. 1-91. Committee members are: Sakurai, Michiharu, chairman (Senshu University), Asada, Takayuki (Tsukuba University), Itoh, Yoshihiko (Seikei University), Ogura, Noboru (Tohoku University), Kobayshi, Noritake (Keio University), Satoh, Yasuo (Housei University), Tsuji, Masao (Waseda University), Hiromoto, Toshiro (Hitotsubashi University), Matsuda, Shuichi (Waseda University), Miyoguchi, Shuji, (Yokohama National University), Monden, Yasuhiro (Tsukuba University), Itoh, Kazunori (Tamagawa Gakuen University).

CHAPTER 2

The Impact of FA and CIM on Business Management

For management accountants, the most important changes in recent decades have been caused by the rapid movement toward more automated and integrated business methods. These new methods cause problems for traditional managerial accounting forms of analysis because they violate the set of unspoken assumptions on which they rest. For example, most traditional industry accounting methods assume that direct labor is an important cost driver. However, this is not so in factory automation (FA). As a result, some methods lose their primary purpose—for instance, standard costing's aim to control direct labor. In addition to the loss of focal point, other assumptions are also violated—for instance, the implicit assumption that standard costing is useful for control purposes in situations with low to medium amounts of variability, but not in situations with high variability.

To develop a comprehensive understanding of the managerial accounting implications of FA and CIM, we must examine their nature as well as attempt to define them clearly. These analyses reveal both the changes in Japanese business management occurring with the installation of FA and CIM, and how these changes impact managerial accounting.

Management must have a survival strategy if it is to cope with the technological, economic, social, and market changes discussed in Chapter 1. One such strategy is factory automation. Already widely adopted in Japan, FA continues to exert a significant influence on the ways of managing businesses even as these companies rapidly proceed toward CIM. To them, CIM signifies a flexible, integrated system that seeks to achieve the operational efficiency needed to implement an advanced management strategy that links marketing, engineering, and production. Japanese managers who confront advanced manufacturing technology are seeking new methods of business management suited to FA and CIM.

FA History

A brief look at the history of FA will make it easier to understand why it has become a management and accounting issue since the 1980s. Although FA's remarkable development in Japan cannot be traced completely because of the many variables, the following discussion presents some of the primary influences commonly thought to be involved.

In the 1960s, mass production technology (low variety/high volume) made striking progress in Japan. As the technological revolution swept production, process-oriented industries (such as steel and petrochemicals) grew. In contrast, assembly-oriented industries producing a wide variety of products in small lots experienced severe problems in manual production—problems for which they had no suitable solutions. These problems went unnoticed in the 1960s because of the low demand for such products. Consequently, the role of these industries in the Japanese economy was small compared to those employing mass production methods.

The production of a diverse array of products in moderate and small amounts by assembly-oriented industries began to attract attention after the 1970s (Economic Planning Agency, 1989). Two factors led to this. One was the increasing diversity of consumer demand, and the other was the development of FA. On the demand side were the maturation of Japanese consumer society, a diversification of value systems that came with higher standards of living, and the development of luxury products. These increasingly diverse consumer needs became conspicuous after the oil crisis of 1973. From that time onward, the rate of proliferation of durable consumer goods became comparable to that of the developed countries of the West. At the same time, the phenomenon of proliferating consumer needs became apparent to observers. Faced with these changes, companies had to produce a greater variety of products in lower volumes to be competitive.

Historically, it was expensive to produce large varieties of products in small lots. However, the advent of flexible manufacturing systems (FMSs) and computer-aided design/computer-aided manufacturing (CAD/CAM) in the 1970s made such production methods efficient. The reader should distinguish this type of FA from the automation of process-oriented factories that flourished in the 1960s (Yoshikawa 1984). Table 2-1 contrasts FA (Dilts and Russell 1985) with the process automation that flourished in the 1960s.

Table 2-1. Comparison of Typical Process Automation and FA

Factors	Process Automation	FA (Factory Automation)
Variety	Low	High
Production volume	Mass production	Low production
Inventory	High	Low
Sensitivity to demand	Slow	Quick
Use of space	Extensive	Small
Quality	High	Very high

Why did Japan create the massive innovation in FA in the 1980s? Many factors may have contributed. One of the most powerful was certainly the labor shortage, signaled by the achievement of zero population growth in 1975 (i.e., a 2.0 fertility rate) and a continuing, potentially devasting decline—the fertility rate in 1993 was 1.46. In this scenario, FA was more than a business opportunity—it was a survival priority! The need to reduce labor requirements has become more focused as the shortage has expanded up the skill ladder to reveal the scarcity of engineers and skilled workers in the late 1980s (The Japan Machinery Federation 1992).

The impending labor shortage was reinforced by other factors. One was growing resistance to difficult, dirty, and dangerous jobs. Another was the rapid fluctuation in the exchange rate and pressure placed on Japan by Asia's newly industrialized countries. Taking these factors together, both on the supply and demand side, we can see how they may have created the conditions that enabled Japanese managers to justify the investment in what was then a financially risky strategy.

The Elements within FA

First we need to sort out our terms and definitions. Confusion can arise because Japanese and American readers have different understandings of the terms FMS, FA, and CIM. For example, FMS (and more recently, CIM) generally has been used in the United States to describe the automation of factories. In Japan, however, the term FA is the more widely used all-inclusive term. In the United States, FMS was the fashionable term from the mid-1970s to the early 1980s. CIM originated with Joe Harrington back in 1973 (Buckroyd 1989). It has had its present meaning since about 1982, whereas CAD/CAM has been used since around 1972 (Koenig 1990). However, the term FA is also sometimes used in the United States (Vanderspek 1993).

Distinguishable from process automation, FA means the coherent automation of a factory. FA extends to the entire factory including technology, design, and the flow of materials, as well as the flexibility to cope with the production of a variety of products in medium and small volume. Japanese FA is composed of flexible manufacturing systems at its heart, computer-aided design/computer-aided manufacturing/computer-aided engineering (CAD/CAM/CAE), and office automation (OA) or information technology (IT). Figure 2-1 illustrates the FA relationships.

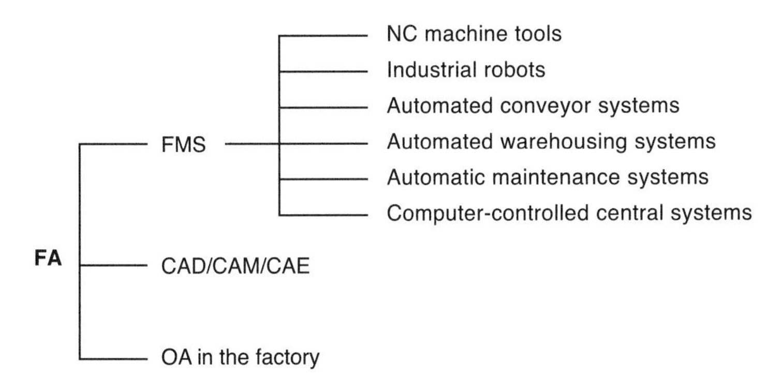

Figure 2-1. What Is FA?

Flexible Manufacturing Systems

Flexible manufacturing systems are the heart of factory automation. FMS allows for the production of wide varieties of products in small lots. It employs numerical-control (NC) machine tools, industrial robots, automated conveyor systems, automated warehousing systems, and automatic maintenance systems. Of course, the creation of each specific manufacturing system differs depending on the type of production and scale. For example, transfer machines and specialized machinery systems are used for low variety/high volume production. However, while appropriate for increasing the productivity of mass-produced products, they lack flexibility.

On the other hand, NC machines are often used in high variety/low volume production because their changeover times (for producing a different item) are quick. However, they have difficulty achieving high volumes. Nowadays, with flexibility more and more necessary, peripheral devices like automatic conveyors and automatic warehouses are arranged around NC machine tools and industrial robots—all managed by a computer-controlled central system. This arrangement tends to allow greater overall productivity and flexibility in the production process.

Computer-Aided Design/Manufacturing/Engineering

Automating a factory does not stop at automating the production system alone. The complexity of many high-tech products has made the engineering drawings more complex—and increased the number of those drawings. Hence, to increase precision, quality, and speed of design, CAD has become necessary. In fact, certain kinds of design (such as the design of integrated circuits) can only be done with CAD. Today's integration of CAD, CAM, and CAE makes it possible to have a coherent design, engineering, and manufacturing system.

Office Automation

Office automation (OA) is a system designed to increase efficiency of office work by (1) reducing the labor devoted to clerical tasks and (2) supporting creative operations. Major Japanese companies gave a lower priority to productivity in office areas vs the manufacturing area. However, OA makes the paperless factory a reality. Through standardization efforts, preparing documents related to production, management accounting operations, communications, and storage operations are more efficient. At the same time, in the support of creative tasks, computer and communications systems now provide communication methods such as teleconferencing, means for accessing pertinent information, and modes for analysis and processing.

Today, thanks to the development of information technology (IT) in PC architectures and software, there has been a movement toward the use of PCs for CIM. Client/server systems have become widespread in CIM environments. Instead of hierarchical models, horizontal open CIM has been widely used in Japanese industries (Inagaki 1994).

The Emergence of CIM

Market diversification and the need for individualized products has become more intense in the 1990s. There is greater use of point-of-sale (POS) technology, a method that links marketing information directly to engineering and production. Conceptually, POS links market information promptly to the development and manufacture of new products. As the market churns with greater frequency—and with smaller orders—Japanese managers seek to reduce delivery times and lead times. To gain competitive superiority, they also seek greater efficiency and higher quality in small-lot production. These forces inevitably lead to "just-in-time" production (also called kanban production) in order to produce the required quantity of products at the exact time when they are needed.

Three Views of CIM Integration

Everyone understands CIM as a computer-oriented, computer-integrated manufacturing system. CIM's importance is its goal of integrating the traditional "islands of automation" (Bray 1988). However, views of what to integrate into the manufacturing system differ and fall into three categories:

1. integration of engineering and production
2. integration of marketing, engineering, and production
3. integration of management, marketing, engineering, and production

The first view (integrating engineering and production) is most often used in the United States. The Japanese typically employ the second view—adding marketing to the equation. The third view—adding management—is more expansive. In practice, it is difficult to clearly distinguish between (2) and (3) because, even when intending to build a system that integrates only marketing, parts of management inevitably creep in. Although many companies are trying to develop them, few implemented CIM systems fully integrate management.

Integrating engineering and production

CIM is most narrowly regarded as an integrated engineering and production system. For example, one argument says that "the essential difference between conventional factory machines and the automated production inherent in CIM is the latter's integration of all information technology required to design, produce, and deliver the product" (Hunt 1989). Another view (Gosse 1989, 1993) defines CIM as "computer support technology ... applied to the manufacturing process that has a regulated reciprocal linkage." The manufacturing process referred to includes CAD/CAM and is considered essentially to be an integrated system of technology and production. Others also limit CIM to the realms of engineering and production (for example, Godher and Jelinek 1985).

In the United States, many companies initiated factory automation by integrating their design and manufacturing areas. Therefore it is natural for some American managers to understand CIM as integrating engineering and production with the intensive use of a computer network and database. However, except for its emphasis on the use of computers and databases, this concept of CIM is what Japanese managers call FA.

Integrating marketing, engineering, and production

The view that CIM integrates marketing, engineering, and production aims for a system that quickly perceives the needs of consumers, links those needs to planning, design, and production, and rapidly fulfills the needs. For example, Koenig (1990) states that excellence in communication is achieved through a common database and that design, manufacturing, and marketing information are the major paths available to a manufacturer. This same reasoning is expressed by Matsushima (1990) of IBM Japan. He argues that CIM "is an information system that integrates the various functions of marketing, engineering, and production under an operating strategy."

Judging from U.S. literature, few Americans favor this view of CIM, while it is the one held by most Japanese managers. The latter are introducing these techniques rapidly in an attempt to strengthen the link between marketing and manufacturing, a concept expressed in the following statement: CIM is an information system that integrates marketing, engineering, and manufacturing using a computer network and database with business strategy at its core (*The Nikkei* 1990).

For example, Snow Brand Milk Products, Japan's largest manufacturer of dairy products, rationalized operations using this type of CIM and thereby increased the efficiency of the flow of materials. The main purpose was to enhance coordination between marketing and manufacturing. In another example, Honda Motors emphasized the overall management of manufacturing and marketing by developing a companywide network that includes manufacturing subsidiaries (FA Report 1992).

A number of CIM research studies revealed that this form of CIM that integrates marketing with engineering and manufacturing is the most common form in Japan. For example, the 1987 survey conducted by the Committee on the Status Quo and Future Outlook for Promoting FA shows this view held by 51 percent of respondents. Sakurai and Huang's survey (1988) also supported this result.

Integrating management, marketing, engineering, and production

This view expands CIM to include not only engineering, marketing, and production but also management. For example,

Groover (1987) explains that "the CIM concept is one in which all of the company's activities related to the production function aid in operational activities, complement them, and are included in an integrated computer system to achieve automation." He lists handling orders, accounting, payroll, and managing credit sales as operating management functions. Buckroyd (1989) likewise gives CIM a broader base than FMS because it is specifically aimed at the business level. He likens CIM to an umbrella encompassing material requirements planning (MRP), FMS, CAE, and computer-integrated testing (CIT).

In Japan, Hitachi's CIM system integrated management with engineering and production. CIM differs from FA in this company in that "to make FA the core of the system, a management system, a marketing information system, a management evaluation system, a personnel management system, and a financial system have been added, and the whole of it has been made into an on-line system" (Tokunaga and Sugimoto 1990). As a result of CIM, the number of personnel involved in indirect tasks on the assembly line, and white-collar workers involved in clerical and managerial areas have been reduced.

The goal of true CIM installation is to completely integrate all components of the enterprise into a single unified system. However, CIM's higher aim is to improve the company's profitability (Koenig, 1990). For some companies it may not be cost-effective to strive for complete integration. In fact, for all practical purposes, such complete integration has yet to be achieved in the United States (Vanderspek 1993).

Similarly, the typical Japanese company has not proceeded fully to "true" CIM, although the numbers are rising. In a 1987 survey only 16 percent of the reporting companies attempted to follow this complete concept. However, in a more recent survey with 236 responding companies, 33 percent wanted to integrate management with manufacturing and technology, 24 percent to integrate marketing with manufacturing and technology, and 24 percent to integrate technology with manufacturing (FA Report 1994).

Mixed CIM installation practices can occur when companies intending only to integrate marketing into engineering and production end up integrating management as well. One example is found in NSK, Japan's largest manufacturer of bearings. Their division in Fukushima strove to integrate marketing into engineering and production. It now produces 10 million bearings monthly with only 350 employees, of which only seven are support staff not directly related to shopfloor production. There is only one cost accountant for the entire division. Such thorough rationalization was made possible because overall management was linked and integrated by a computer network.

How to Install CIM

CIM can be built in two ways. One approach begins with automating a production system and later consolidating these "islands of automation." The other—a top-down method—starts by reforming management. This latter approach generally consolidates the islands of engineering and production before striving to integrate management itself. Compared to Japanese companies, firms run by foreign capital in Japan typically take the top-down approach and ultimately aim to reform management. For example, IBM Japan installed global CIM from the top down. In contrast, while they recognize that the top-down approach is appropriate for CIM conversion, Japanese managers often begin with the aim of harmonizing from the bottom up.

CIM and Strategic Information Systems

According to typical Japanese understanding of management strategy, CIM is considered to be an information system that integrates marketing, engineering, and production via computer networks. An indispensable element is the use of databases (MITI 1989; Yui 1990).

The U.S. views of CIM espoused by Koenig (1990), Groover (1987), Buckroyd (1989), and Hunt (1989) and the predominant view held in Japan are depicted for reference in Figure 2-2. In this book, CIM is considered to be FA integrated by computer.

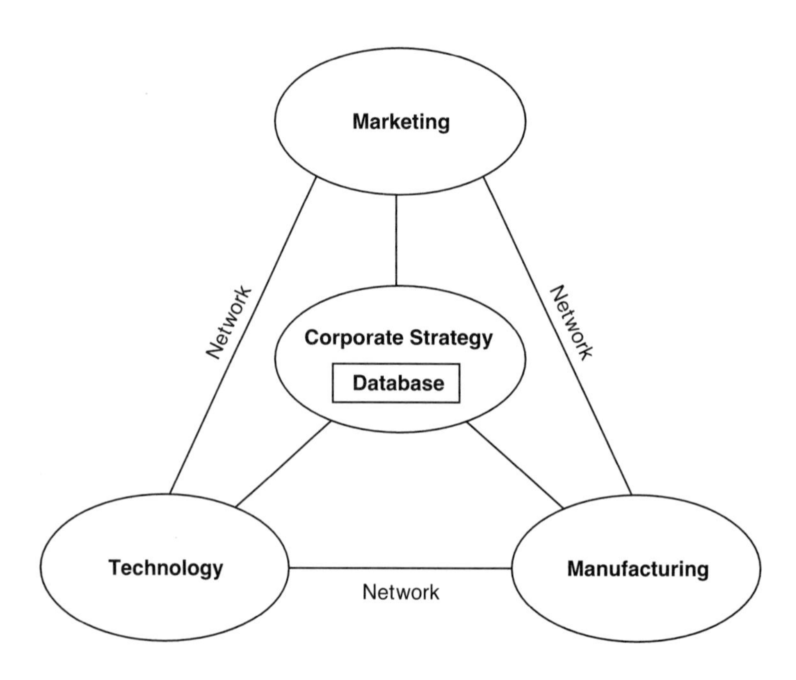

Figure 2-2. Conceptual Diagram of CIM

Regarding the role played by management strategy in Figure 2-2, strategic information systems (SISs) for the 1990s are so important that they hold the key to a company's survival. The strategic application of information links information systems to management strategy and endeavors to establish and maintain superior competitive advantage. For example, Kokuyo, Japan's leading manufacturer of paper stationery, linked wholesalers and retailers to companywide distribution centers. By successfully distinguishing itself for its responsive order-handling network, Kokuyo has maintained its superior competitive position.

SISs are significant in the following ways:

- They differentiate products and services.

- They expand and strengthen relationships with clients and suppliers.
- They strengthen a company's ability to cope with the environment.
- They endeavor to establish and maintain superior competitive positions by advancing into new operations.

Benefits Gained from Implementing CIM

CIM has an effect on the four areas that achieve superior competitiveness in a new environment: cost, flexibility, quality, and service.

Cost. To establish a competitive price, a company must maintain low costs. Although direct labor costs fell through FA, indirect costs rose significantly. Furthermore, the potential for explosive cost increases exists when tightening the linkages between functional areas—particularly if using traditional methods. CIM strives not only to reduce direct labor costs but also to reduce overhead and materials cost by using innovative software linkages.

Flexibility. Although evaluating flexibility quantitatively is difficult, increased flexibility allows for a quicker response to the marketplace. The production of a variety of products in low volume at a low cost becomes a possibility. Product development speeds up and a company can fit production to market needs.

Quality. CIM can bring a substantial improvement in quality. Product reliability will increase and claims and returns due to defective products will decrease.

Service. Service in this context refers specifically to delivery times and customer service. To improve its quality of service to customers, a company must focus on reducing delivery times while maintaining the rate at which products are delivered (the time for an order to be received and the product developed, designed, and shipped). Reducing delivery times significantly shortens a product's lead time (that is, the time from when materials are ordered to when a product emerges).

The Main Aims of Implementing CIM

According to the survey conducted in 1990 by the Japan Management Association, most companies (74 percent) named lead-time reduction as the primary aim of their CIM implementation. The second most common goal was to strengthen links between production and marketing, marketing and engineering, and production and engineering (58 percent). Many respondents apparently aimed for an integrated marketing and production system that could respond immediately to consumer needs and that would link engineering and production. A third goal was to cope with high variety/low volume production (52 percent). These goals all relate to the flexibility and service mentioned previously—and clearly show this to be one of the most important goals of CIM in Japanese companies.

In contrast, these same managers expressed comparatively little concern for CIM goals of reductions in indirect and direct labor costs (16 percent), work-in-process (16 percent), and material costs (1 percent). There were also few expectations that CIM would contribute to improvements in quality (5 percent).

These survey results might obscure certain issues if taken too literally. Prior to any decision to adopt CIM, a company must come to the realization that CIM is the only cost-effective means to obtain the tight cross-functional linkages indicated by the business strategy. Management's expressed goals would be the same no matter what method was used to obtain the links. Therefore, while cost reduction is one of CIM's primary roles, most managers would not view it as a cost-reduction method.

A more recent survey (FA Report 1994) on benefits gained by introducing CIM shows Japanese managers placing cost reduction third in importance (24 percent) behind lead time (34 percent) and flexibility (30 percent).

CIM's Impact on Business Management

How do automation and computer integration affect business operations? The following sections will focus on five influences that result when companies convert to CIM.

A Coherent and Integrated Management System

Companies using traditional production and sales methods to produce large lots of a small variety of products first established production plans based on the forecasted sales for each product. Planners then attempted to stabilize the usage rate of factory equipment by way of a production budget. Today a CIM system makes it possible to obtain a coherent and integrated linkage between marketing, engineering, and production areas.

By creating a link between marketing and production, customer needs are quickly ascertained and fulfilled. This substantially reduces the cost of design and testing in engineering as well as the learning curve phenomena in production. It now becomes possible to design new products quickly and efficiently. Endowed with flexibility, a factory can produce different types of products in the desired amounts. As Porter writes in his book *Competitive Advantage* (1985), by forming a value chain of activities from sales to engineering and production, a company can reduce costs and maintain its competitive advantage.

Horizontal, Not Functional Organization

In the past it was enough for individual employees to carry out the orders of their superiors without understanding the overall goals of the company. If shopfloor workers faithfully performed their jobs, it was presumed that the company could achieve its goals. This was the stage of the pyramid-shaped functional organization.

In the CIM environment, a company responds to changing customer needs through flexible production. This makes good communication essential between engineering and management. It is crucial that manufacturing activities satisfy marketing needs—not the other way around. The key is maintaining a flexibility that quickly recognizes market trends. Under such circumstances, the traditional military-type organization with its emphasis on command and control no longer suffices. An organization that has adopted a participatory network style (Koenig 1990) is more adaptive.

A typical example of Japanese organization is the *Syukan* (general manager) organizational system at Nissan. Devised to achieve target costing, the Syukan is responsible for a product from product planning to marketing; its organization is horizontal, not functional.

Reduction of Direct Labor Costs

The introduction of industrial robots reduces direct labor in the workplace. Toshiba Tungalloy, Japan's largest manufacturer of super-hard tools, reduced the number of worker-operators from 70 to 16 by introducing FMS. Niigata Engineering, a midsized machinery maker, reduced workforce numbers from 31 in the traditional process to 4 in the new process. Yamazaki Steel Fabricating reduced the number of employees in direct operations from 204 to 10 (The Economic Research Institute 1982). Typically, workers who found their jobs eliminated in one place were rotated to other shop floors. Some studies make it clear that, while there is a difference in degree, increased productivity due to workforce reductions in direct operations is not particular to specific companies (JMA Research Institute 1983; Yoshikawa 1984).

Employees directly involved in traditional operations gradually disappear from the workplace to be replaced by monitoring, maintenance, R&D, and software development personnel. Although direct labor costs decrease under FA, indirect labor costs tend to rise. In contrast, because CIM's focus is on reducing planning, design, and paperwork, we can expect to see a relative reduction in personnel involved in indirect tasks.

Scientific Methodology

The increasing use of machinery and equipment in the workplace will continue to replace human beings, thereby reducing human error. The more consistent work of industrial robots will result in greater precision of operations and greater reliability.

Traditionally, companies relied on unscientific methods—for instance, making estimates based on past experience. As a result of CIM and the resulting "superiority in communication," management style is more scientific. This makes decision making more precise than in the past (Koenig 1990).

However, typical Japanese managers do not believe in conducting management with minute calculations based on pure theory. They feel that management in the future will require even more human judgment. So while Japanese managers do not believe that actual management will change as simply as Koenig argues, it is likely that CIM will result in management becoming more detailed and sophisticated.

Growing Impact of Computer Software

The use of computer software—and so also, information investment—increases with CIM. These computer costs are so enormous that they will exceed the cost of tangible assets in the next decade. This means that accounting and cost management for software and information investment will grow in importance to management—perhaps becoming the most important single cost component in the near future. (Managerial accounting issues related to software are discussed in Chapters 9 and 10.)

Conclusion

This chapter has presented the history of FA and CIM, discussed the benefits of their introduction, and touched on CIM's impact on business management. Forthcoming chapters will discuss how management accounting was or should be changed in the advanced manufacturing environment. Chapter 3 will open the discussion on target costing for strategic cost management.

References and Further Readings

Bray, Olin H. 1988. *CIM Computer-Integrated Manufacturing: The Data Management Strategy.* Bedford MA: Digital Press, p. 18.

Buckroyd, Allen. 1989. *Computer-Integrated Testing.* New York: John Wiley & Sons, pp. 3,4.

Dilts, David M., and Grant W. Russell. 1985. "Accounting for the Factory of the Future." *Management Accounting* (April) p. 37.

Economic Planning Agency. 1989. *Position Paper on the Economy: The Beginning of Heisei Era Economics and New Currents in Japan's Economy.* Finance Ministry Printing Agency (August 30) p. 80.

The Economic Research Institute, Japan Society for the Promotion of Machine Industry. 1982. *An International Comparison of Management in the High Technology Industry* (March) pp. 59, 92.

FA Report. 1992. *Magazine for FA and CIM* (October) pp. 22-25.

FA Report. 1994. "Analysis of Mail Survey on CIM." *Magazine for FA and CIM* (February) p. 10.

Goldhar, J. D., and M. Jelinek. 1985. "Computer-Integrated Flexible Manufacturing: Organizational, Economic, and Strategic Implications." *Interfaces* (May-June) pp. 94-105.

Gosse, Darrel Irvin. 1989. *An Empirical Field Study of the Role of Cost Accounting in a Computer-Integrated Manufacturing Environment.* Ph.D. Dissertation, Michigan State University, p. 12.

Gosse, Darrel Irvin. 1993. "Cost Accounting's Role in Computer-Integrated Manufacturing: An Empirical Field Study." *JMAR* (fall) p. 59. The definition of CIM is different but on the same line as in his earlier dissertation.

Groover, Mikell P. 1987. *Automation, Production Systems and Computer-integrated Manufacturing.* Englewood Cliffs: NJ: Prentice-Hall, pp. 771, 772.

Hunt, V. Daniel. 1989. *Computer-Integrated Manufacturing Handbook*. Chapman and Hall, pp. 45, 239.

Inagaki, Kunihiko. 1994. "Downsizing and FA Network." *FA Report, Magazine for FA & CIM*. (July) pp. 17-18.

JMA Research Institute. 1983. *Robotization: Its Implications for Management*. Fuji Corporation, pp. 116, 126, 171, 214, and others.

Japan Machinery Federation. 1992. "FA Trends: Survey Report for Compiling Statistics." *International Robot*/Factory Automation Technology Center (June) p. 49.

Japan Management Association. 1990. *Management Issues Survey (Production)*. Japan Management Association (November). In this CIM survey, production managers of 800 manufacturing companies listed in Section One of the Tokyo Stock Exchange were contacted and 282 responded.

Koenig, Daniel T. 1990. *Computer-Integrated Manufacturing: Theory and Practice*. New York: Hemisphere Publishing Corporation, pp. 3, 5, 15.

Matsushima, Keiju. 1990. *Manufacturing Transformed by CIM*. Kigyochosakai, p. 24.

MITI Information Agency. 1989. "The Outlook for FA: From FA to IMS." *International Robot*/FA Technology Center (July) pp. 5-6.

The Nikkei. 1990. "Factory Transformation: Aiming for Higher Productivity through Information Networks" Nihon Keizei Shinbunsha Inc. (September 12, Morning ed.) p. 7.

Porter, Michael E. 1985. *Competitive Advantage: Creating and Sustaining Superior Performance*. New York: The Free Press, pp. 33-53. (Translated by Toki Mamoru, Nakatsuji Manji, and Onodera Takeo. *Strategy for Competitive Superiority*. Diamond Inc., pp. 45-68.)

Sakurai, Michiharu, and Philip Y. Huang. 1988. "Factory Automation and Management Accounting Systems." *Business Review of Senshu University* (September) p. 74.

Tokunaga, Shigeyoshi, and Noriyuki Sugimoto. 1990. *From FA to CIM: A Study of the Hitachi Experience*. Dobunkan Shuppan, p. 14.

Vanderspek, Peter G. 1993. *Planning for Factory Automation: A Management Guide to World-Class Manufacturing*. New York: McGraw-Hill, pp. 70,75-80.

Yui, Shigetomo. 1990. *CIM: Achieving Integration of Production and Sales*. Nihon Keizai Shinbunsha Inc., pp. 139-156.

Yoshikawa, Hiroyuki, ed. 1984. *Recent Examples of FA of Factories*. Techno Publishers Co. Ltd., pp. 515-1,081.

CHAPTER 3

Target Costing for Strategic Cost Management

Target costing is a comprehensive means of strategic cost management focusing on total life cycle cost reduction. To achieve life cycle cost reduction, target costing integrates production and marketing functions with engineering as the core discipline. Independently, the CIM environment integrates production, technology, and marketing functions with a communications network to improve their interaction. Since CIM and target costing both involve a similar integration, they are often used together. In fact, due to enhanced communications, target costing is considered most powerful when used in a CIM environment.

The spread of target costing in Japan is driven by many factors. These include diversifying consumer needs, which lead to shortened product life cycles and intensified international competition. With products designed—and redesigned—more frequently, cost reduction efforts must focus on the design process.

Target costing has also spread to Germany (Horváth 1993; Seidenschwarz 1993), the United States, and other Western countries (The Society of Management Accountants of Canada 1994). While target costing has become a standard practice in assembly-oriented industries, modified target costing systems have also been developed for process-oriented industries and for computer software companies.

This chapter focuses on target costing for strategic cost management in the high technology environment.

A Short History of Target Costing

Following the early 1970s, target costing spread quickly in Japan primarily to assembly-oriented industries. In a survey of 270 companies listed on the Tokyo Stock Exchange, 52.4 percent responded. Of these respondents, 84 percent had introduced target costing (Sakurai 1988). The industries surveyed were largely high technology manufacturers: precision machinery, electrical equipment, transportation equipment, machinery, metal products, and miscellaneous manufacturing. A later survey made by Kobe University Researcher's Group (1992) supported the above findings.

Development of Target Costing

In the 1960s, process-oriented industries such as steel companies and large petrochemical firms grew rapidly due to the development of process automation. Demand for products in that era was strong. It was often called the "era of new products" or "the era of fulfillment" because for the first time in Japanese history consumers had the purchasing power to drive the market. This period of remarkable social and economic transformations conditioned the present environment in Japan.

At that time, mass production of standardized products was the major production form in Japanese companies. The role of planning and design in the production process was not considered an important focus of cost management. In this low variety/

large lot production environment, cost management focused on the production process. Thus, standard costing was the major means of cost control until the early 1960s.

Japan's standard of living rose significantly in the late 1960s and the 1970s and consumer consciousness diversified. The market overflowed with products to such an extent that it was called a "period of saturation." Afterward, the proliferation rate of durable consumer goods reached the same level as the most highly developed countries. Companies had to produce a large variety of products with distinct characteristics in order to respond to this diverse demand. The systematic application of computers began in the late 1970s. The dramatic penetration of industrial robots and NC machine tools gave companies the ability to efficiently produce a large variety of products in small quantities (Hasegawa 1984). We now recognize this period as the beginning of large-scale factory automation.

Product life cycles became shorter as consumers searched constantly for newer and "better" products. Shortened life cycles have naturally increased the importance of cost management in the planning and design stages (Imai 1987). This is because (1) the preproduction stages determine the cost structure, and (2) there is no "long run" in which to reduce costs. The importance to manufacturing of these earlier stages is shown in Figure 3-1.

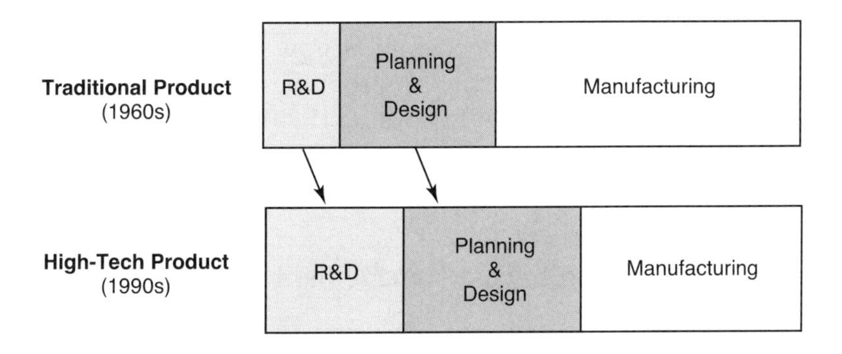

Figure 3-1. Life Cycle Differences Between Traditional and High-Tech Products

Target-costing activities prior to the 1973 oil crisis generally took the form of value engineering (VE). Instances can be found of Toyota using target costing as early as 1963, and in 1966 Nissan utilized a cost management program for new car development similar to today's target costing. However, target costing in its present form really spread following the oil crisis. Leading companies such as Daihatsu Motors employed a framework for present target-costing activities around 1974 (Kato 1990).

Target Costing Today

As FA and CIM reduced the number of workers on the shopfloor, standard costing became less important as a means of cost management in assembly-oriented industries. Consequently, the key issue in Japanese companies became how to reduce costs at the planning and design stages. Clarion's Mr. Yamada told the author that cost reduction activities at the planning and design stages had become crucial to company survival in these competitive times because nearly 90 percent of costs were determined at the planning stage.

Since the late 1980s, target costing has become closely connected with business strategy and is considered a strategic cost management tool for attaining the target profit specified by mid-range business planning. For example, Nissan bases target cost on target profit which, in turn, is based on business strategy and customer needs. Thus, target costing is now considered to be a strategic cost management tool for profit planning as well as cost reduction. Recent works such as Kato (1993) and the report prepared by JAA Special Committee (1994) stress the strategic cost management purpose of target costing.

Japanese Approach to Cost Management

Target costing fits nicely with the general approach most Japanese companies take toward operational and strategic cost management. Management effort in Japan generally focuses on three areas: innovation, kaizen (or continuous improvement), and maintenance. Innovation means drastic change brought about as a result of introducing new technologies and/or investing in plant and equipment. Kaizen includes small daily continuous improvements as well as changes in management structure. Maintenance means activities for maintaining current standards in technology, business, and operations. The cost management framework of this focus utilizes target costing for innovation, kaizen costing for kaizen, and cost maintenance for maintenance.

Target Costing

Target costing is a strategic cost management process for reducing total costs at the product planning and design stages. It does this by concentrating the integrated efforts of all related departments of a company such as marketing, engineering, production, and accounting. This process of cost reduction is conducted at the upstream stages of production. As a result of target costing, innovation is promoted.

Kaizen Costing

Kaizen costing involves (1) cost reduction activities for each product, and (2) cost reduction activities for each period. Generally, direct material costs and direct labor costs are controlled through VE (and other engineering activities) and by standard costing for each product. In contrast, overhead is managed primarily through budgeting and by tapping employee know-how via employee-involvement methodologies such as total quality control (TQC) and total productive maintenance (TPM).

Japanese managers attempt to reduce labor hours and materials by detecting unprofitable products and by reducing costs of existing and new products. To reduce cost on a per-unit basis, activities are aimed at reducing the parts, materials, and energy used. The number of employees is reduced, material costs are controlled, and water and gas usage is cut. For example, assume that company X wants to reduce its direct labor costs. There are currently 10 employees assigned to the conveyor line of process A. With a cycle time of one minute, the total labor hours for each unit is 10 minutes. By improving operations with JIT's kanban system, they decreased their labor needs by one operator and reduced the labor hours to 9 minutes per unit.

Kaizen costing necessitates standardizing products and parts, applying VE to all items purchased, and increasing the effectiveness of both equipment use and indirect costs. Figure 3-2 shows a schematic diagram of the relationship between target costing and kaizen costing at company X.

Cost Maintenance

Cost maintenance was essential over the last twenty years. However, the advent of the CIM environment has diminished its importance because industrial robots can produce quality products with low cost.

Cost maintenance means operating at current standard costs for technology, business, and operations. Specifically, it means setting price and quantity standards for product costs and then insuring that actual results are close to the standards. For new products, cost maintenance means attaining the target costs set by target costing. In existing products, the role of standard costing is to reach and stabilize the standard cost.

Cost maintenance activities are cost control activities undertaken to control departmental costs, productivity, unit price, and equipment. In controlling departmental costs, operations improvement is sought by comparing budgeted and actual costs. Controlling productivity involves focusing improvement efforts on reducing labor-hours and on improving value-added productivity. Controlling the unit price of purchased items involves creating a cost table for use when negotiating to purchase parts from internal and external suppliers. Because such control is very important, VE teams are organized to conduct comprehensive cost reduction activities for purchased parts. Unit costs are

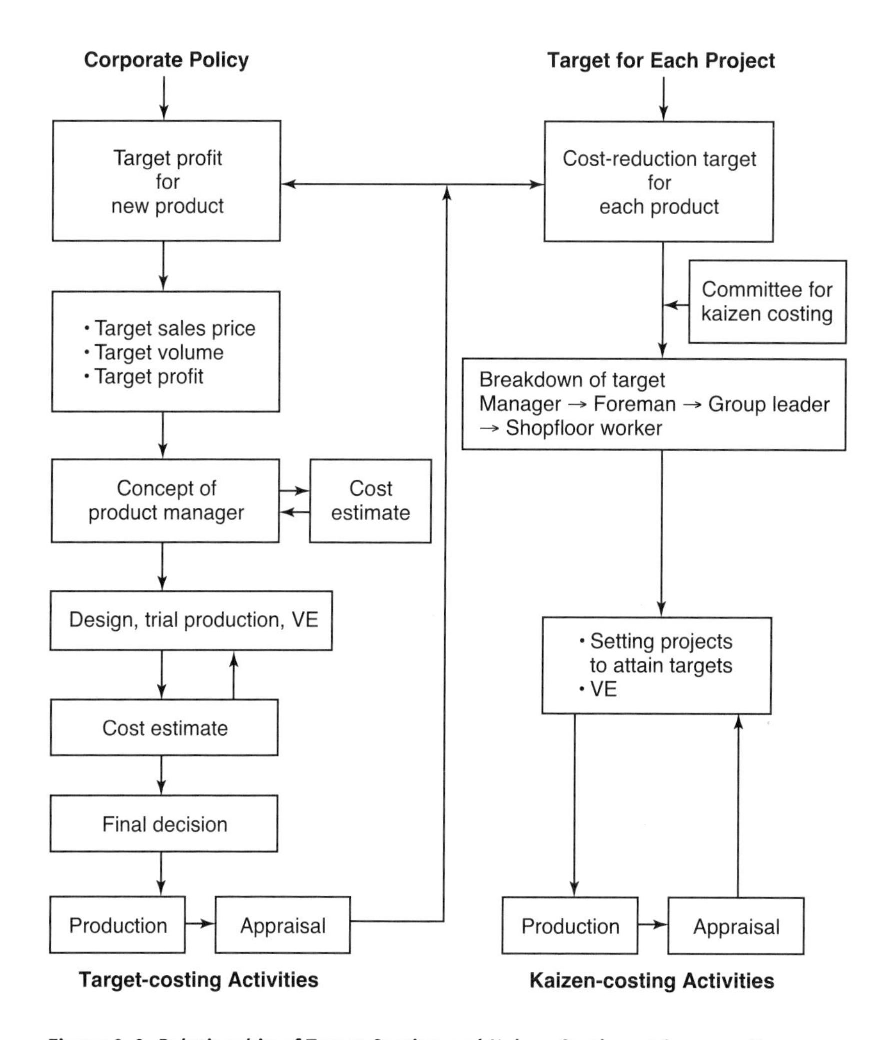

Figure 3-2. Relationship of Target Costing and Kaizen Costing at Company X

revised regularly (normally once a year). Lastly, with respect to managing equipment, particular attention is paid to managing start-up operations in order to improve capacity utilization. By managing equipment budgets, equipment costs can be controlled appropriately.

Goals and Characteristics of Target Costing

The main goal of target costing is to reduce total costs while maintaining high quality. However, many Japanese companies also use target costing for strategic profit planning. We can define these two goals of target costing as follows:

1. *Cost reduction* means reducing total costs (including manufacturing, marketing, and user costs) while maintaining high quality.
2. *Strategic profit planning* means formulating strategic profit plans by integrating marketing information with engineering and production factors.

Specific reasons for introducing target costing differ from company to company. For example, a new Daihatsu division's goal was to have an efficient factory. Target costing at Daihatsu was closely related with factory automation: (1) to realize a factory that emphasizes a cost strategy for establishing profitability, and (2) to advance the efficient use of automation and flexibility. On the other hand, Isuzu's primary aim since beginning in 1966 (it was close to VE at that time) was to reduce variable costs. The chief purpose of the 60 people in Isuzu's target-costing department today is to manage variable costs throughout the company including manufacturing and marketing. Target costing at Canon was initiated in 1980 for the purpose of developing low-cost quality products that were still functionally superior to competing products (Kan 1991).

Seiemon Inaba, chairman of Fanuc, says the goals of new product development at his company are increased reliability, cost reduction, fewer parts, and good design. The tool employed for attaining these goals at Fanuc is target costing (Inaba 1991). The aim of target costing at Fuji Xerox is to attain business profit (Kuramochi 1991). Nissan also considers target costing to be a tool for attaining target profit (Kimura 1992).

Characteristics

Companies differ in their perceptions of what to do with target costing. This in turn affects the characteristics of target costing as a tool. Target costing and standard costing are both tools for cost management. Thus, we will discuss target costing in relation to standard costing as follows.

1. *Target costing and the planning and design stages.* Target costing and standard costing apply at different stages in a product's life cycle. Whereas standard costing is used in the manufacturing stage, target costing is not (see Figure 3-3).

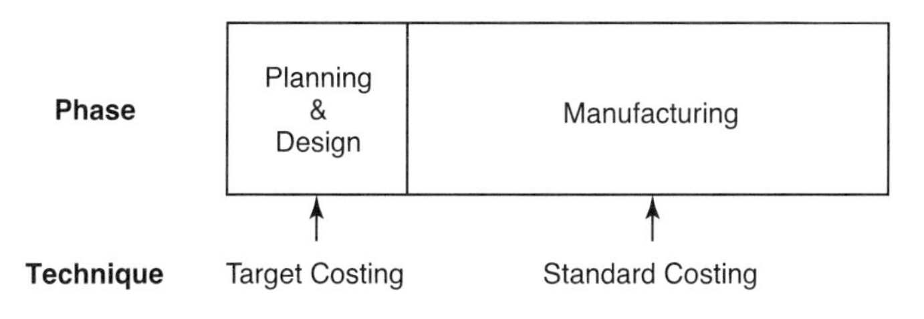

Figure 3-3. Target Costing Use

2. *Target costing is a tool for cost reduction.* Conceptually, cost management can be divided into two parts: (1) cost reduction (or cost planning) and (2) cost control. Standard costing is mainly useful for cost control. On the other hand, target costing is a tool for cost planning. (In fact, target costing is sometimes translated from the Japanese as "cost planning.")

3. *Target costing is a market-driven technique.* Whereas standard costing is driven mainly by production and technology, target costing is driven by the market. Of course, target costing could be implemented exclusively as a technology-driven technique. However, users would not reap the maximum benefits since a tight connection to strategic policy is crucial to the method. This relationship is shown in Figure 3-4.

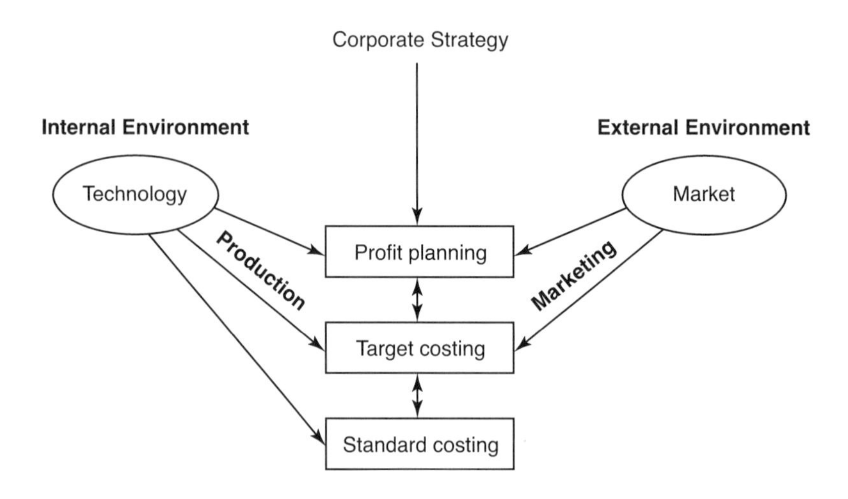

Figure 3-4. Target Costing versus Standard Costing

4. ***Target costing is a part of strategic profit planning.*** In target costing, it is assumed that the strategic business plan is formulated by considering competitiveness and customer needs. In fact, target costing is often used as a bottom-up tool to attain the target profit set by top managers determining corporate strategy. Thus, the cost reduction program is bound tightly to the target profit set at the strategic level. In contrast, standard costing is an operation-level cost-control tool usually based on nonstrategic engineering assumptions.

5. ***Target costing is an engineering-oriented technique.*** Target costing is a management tool for directing and focusing the decision process for design specifications and production engineering. Financial accounting measurements are not emphasized in target costing, reflecting more engineering characteristics. It harmonizes with other Japanese management engineering techniques such as VE, TQC, and JIT.

6. ***Target costing depends on and enforces cooperation between departments.*** In target costing, the accounting department

acts as coordinator and information provider while the marketing, engineering (planning/design), and production departments determine success or failure. At this point we see the convergence of target costing and CIM and the resulting integration of marketing, engineering, and production.

7. Target costing works better in high variety/low volume production. Standard costing is most effective when applied to standardized mass production. In contrast, target costing is not well-suited for mass production. Also, current versions work better in assembly-oriented industries than in process industries.

Target Costing in Manufacturing Companies

This section describes the target costing process in some detail. In addition, we discuss implementation issues including organizational characteristics that assist or impede success; alternate methods and scope for the procedure; and integration of the organizational structure, target costing, and strategic planning.

Target Costing Procedure

There are three primary conceptual steps in target costing, although companies develop and customize their own specific procedures (Nobbori and Monden 1983; Monden 1991).

Step 1. Plan new products by focusing on customer satisfaction.

Step 2. Determine the target cost through company strategic policy and align it with achievable costs.

Step 3. Attain the target cost by using VE or other cost reduction techniques.

The target cost is determined in step 2. When there is a request from a customer, the allowable cost is calculated by subtracting the target profit from the planned sales price. The allowable cost is known as the "maximum permissible manufacturing cost"

(Tanaka 1979). It is not based in cost accounting at this point. Rather it is the estimated cost based on market conditions.

The next step is to figure out if the product can be made for this amount. The "drifting cost" is calculated for each part based on accounting records. This drifting cost is also called an estimated or base cost and is a current estimated cumulative cost with no target in mind. It is referred to as drifting because it is recalculated continuously as VE work is performed. In fact, the primary work in target costing is the VE effort to reduce the drifting cost until it equals the allowable cost.

The index used to set the target profit is usually return on sales (ROS) rather than return on investment (ROI). One reason may be the ease of calculating the ROS for each product. However, more importantly, the use of ROS is strategically superior to ROI. (This will be discussed in Chapter 11.)

The target cost is usually attainable only through a painstaking VE program. Because the allowable cost is usually the desired cost indicated by top management, it tends to be rigid. As a result, if the drifting cost does not reach the goal, then additional cost reduction activities are carried out with VE programs for second and third estimates. Finally, a target cost is established that is attainable and can also be the target for production efforts. To make the target cost a challenge, some companies require that the allowable cost be equal to the target cost.

The target cost is determined through profit planning (via ROS) while coordinating with the drifting and allowable costs. It is set as a challenging target for each new product arrived at through the interaction of profit planning with technical planning. Target costs set in this way are finally approved and authorized by top management.

Plant foremen normally are responsible for additional cost reduction programs once the target cost is determined. Generally, the learning curve effect and/or cost reduction through volume production are incorporated in setting the standard cost. The challenge is now to attain the newly set standard cost.

Following an adjustment period, a performance report is examined to check whether any cost standards have not been achieved. If any abnormalities have occurred, the problems are brought out in "kaizen costing committees" that discuss proposals for improvement. This means target costing is also a cost reduction activity based on self-imposed factory improvement efforts. An indispensable condition for its success is total employee involvement in cost reduction activities.

Organizational Structure for Promoting Target Costing

Target costing should be made a companywide program that involves the expertise of everyone from the product development and design stages through manufacturing to volume production. This makes it imperative to determine the target cost reasonably and to perform follow-ups for each step of target costing.

Different companies propose various ways to proceed from development and design to volume production. While there is no set way to do this, the method adopted by one company is shown in Figure 3-5.

New product development must necessarily focus on satisfying customer needs. In addition, a new product should be competitive in terms of price, be reliable, have superior features, be user friendly, have superior durability, and be accompanied by good service. At this stage, quality cost is considered and a competitive product is created.

Headed by the product development manager, the product development committee reviews the new product with representatives of product development, planning/design, manufacturing/engineering, and purchasing. The product development process proceeds through its various stages with progress reviewed at each stage by the target costing team. In addition, the evaluation results are discussed at division and companywide levels. Following approval, a project proceeds to the next step.

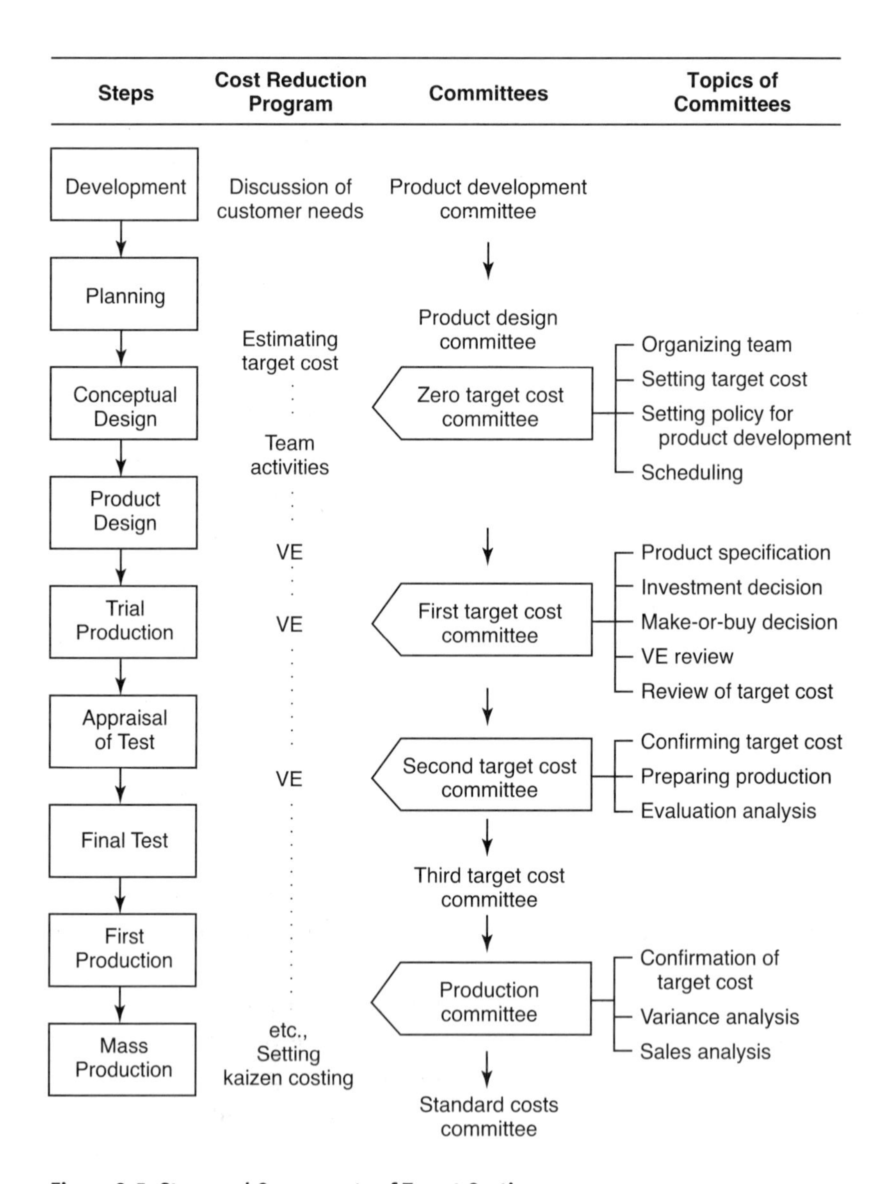

Figure 3-5. Steps and Components of Target Costing

Determining Target Cost

Determining target costs is extremely important because it is tied to the company's strategic policy. Also, the reasonableness of the targets will determine employee reaction. Management accounting is involved at many points in the process of setting target costs. There are basically three methods for determining target cost—the profit planning method, the engineering method, and the combination method. These methods will be discussed in turn.

Profit planning method (top-down)

The profit planning method is also called the apportionment method. It determines the target cost by subtracting the target profit from the projected sales price. Target cost is an estimated total cost for insuring target profit after considering the sales price of competitors and other factors. It is a top-down method; according to Tanaka (1987), this method was used by around half of Japan's companies.

Engineering method (bottom-up)

The engineering method is also called the cumulative method because it sums up (accumulates) all costs based on current methods. It determinines the target cost by considering the present level of technology, production facilities, delivery time, production volume, and others based on the company's technological level. This is more a bottom-up method because the target cost is determined by operating engineers using engineering methods—not by top management.

Combination method

The combination method consolidates the operating focus on profitability and the technological focus on feasibility. The target cost is determined by integrating the other two methods.

All three approaches are practiced in Japan. However, the combination method is preferred because it leads to coordination between marketing, engineering, production, and accounting departments (Sakurai 1988). In fact, a survey indicates that target costing in Japanese companies does not rely solely on either the profit planning or the engineering method but rather on the combination method (Sakurai 1991). Many companies switched from other methods to the combination method.

A 1992 survey conducted by Kobayashi supports the earlier survey. It indicates that 57 percent of 106 Japanese companies listed on the Tokyo Stock Exchange use the combination method (Kobayashi 1992). The case presented later in this chapter is based on the combination method.

Costs Included in Target Cost

Full absorption cost is the most common cost category used in target costing. When the target cost includes manufacturing costs, they are typically categorized as direct material costs (classified further as purchased parts costs, internally manufactured parts costs, and cost of dies), direct conversion costs (direct labor costs, direct equipment costs, and subcontract costs), and indirect conversion costs (indirect labor costs, indirect equipment costs, and others).

The primary objects of cost reduction in target costing are direct material costs and direct conversion costs. Because of this focus, some companies include only these costs in the target cost. For example, a manufacturer of electric generators breaks down target costing's primary cost reduction areas into direct material costs and direct conversion costs. It treats development and design costs as direct conversion costs and computes the target cost for development and design by multiplying estimated labor-hours by a standard rate.

In another example, a large automobile manufacturer treats parts costs and material costs as direct material costs. It treats direct labor costs, depreciation costs of dies, machinery, and other direct costs as direct conversion costs. It also includes development costs in other direct conversion costs because they are the cost item that can and should be managed by the product manager. All of these costs are the objects of target costing. Management emphasizes conversion costs as well as direct material costs as the most important item in determining target cost. The company controller explains: "Manufacturing processes are classified as separate processes, so there is a manager for each process. And because the organizational sphere of management is clearly defined, it is easy to manage these conversion costs."

Some companies—Isuzu, for instance—make only parts and materials costs that are the object of VE the objects of target costing. On the other hand, another automobile manufacturer includes all variable costs in the target cost and uses marginal profits as target profits. If fixed manufacturing costs and marketing costs/general administrative costs are included in the target cost, they must be controlled. However, fixed manufacturing costs and operating expenses are difficult to control. Hence, this automaker puts only parts costs and variable costs within the scope of target costing.

When developing products that are similar, it may be better to limit the object of the target cost to the differential costs. For example, when Toyota makes a model change, the cost of the existing model being produced is fixed. By computing only the cost differences for the new model, managers are able to grasp the cost variations that originate from design changes and volume changes and thereby ignore other variations in costs. This has been called the "differential estimate" (Tanaka 1990). This method not only reduces the effort of making an estimate, it has the advantage of increasing accuracy and making it easier for engineers to understand variations in costs.

In new product development, the allowable cost should be calculated on the basis of total absorption costs that include marketing costs and general administrative costs, not just manufacturing costs. That practice is followed because the full cost principle is very often used in pricing and profit planning. As a reference, one should think about the allowable cost from the perspective of life cycle costing. In other words, user costs such as operation, maintenance, and disposal should also be considered as part of the allowable cost.

However, in setting the drifting cost, one should focus only on controllable costs such as variable costs or manufacturing costs because they are the major concern of cost reduction on the shopfloor. Major engineering concerns inevitably will be direct material costs and direct conversion costs. As discussed previously, this may be why many companies include only these VE cost items in the target cost. Although uncontrollable costs are often included in determining the drifting cost, they should be identified separately for cost reduction.

Finally, some companies use the factory invoice price. For example, Fuji Electric subtracts the target profit from the factory invoice price and then calculates the total cost per item. In this situation factory invoice price should be treated as if it were the sales price.

Cost Tables

Cost tables are indisensable for companies that use target costing. As tools for easily and accurately estimating costs for materials, parts, utilities, conversion, and so on, cost tables can be either accurate or rough. Different kinds of cost tables are used for evaluating performance, for effectively purchasing parts and material, and for defining manufacturing methods and their corresponding costs.

Originally developed and used to estimate the price of purchased parts, cost tables are now used to accurately estimate production costs and profit on purchased parts. They are also used as data to show in-house manufacturing costs. Nippondenso, Japan's top producer of electronics and electrical parts, uses cost tables to determine standard costs by estimating such factors as material requirements, conversion processes, plants and equipment, labor-hours, and conversion cost rate in an orderly method (Nishiguchi 1989).

Today Japanese companies are computerizing cost tables to create a more efficient and sophisticated tool. For example, Nippondenso has installed 22 types of cost tables in its sophisticated computer system.

The VE Program

Value engineering (VE) is an indispensable tool for target costing. However, it is defined and practiced differently in Japan and the United States. Also, VE's use in target costing differs from its purpose in other areas.

VE in Japan

VE is a method for doing systematic research on each function of a product or service to learn how to attain required functions with the lowest total cost. In other words, it is a method or tool for reengineering the functions or purposes of a product or service in order to improve its quality or value and achieve customer satisfaction with the lowest cost.

VE can be applied to manufacturing as well as service industries. In target costing, VE is one of the keys to effective new product development.

Types of VE

VE activities can be divided into three categories:

- zero-look VE
- first-look VE
- second-look VE

Zero-look VE (0 Look VE) is applied at the product-planning stage. It is applied to the process of determining what new product to manufacture. This stage opens the door to innovative ideas—to the great benefit of the company. Zero-look VE is sometimes referred to as marketing VE.

First-look VE (1st Look VE) is applied to the development and design stages. Here, VE focuses on the shopfloor and the efficiency of manufacturing activities. First-look VE actions are closely related to manufacturability.

Second-look VE (2nd Look VE) is applied to the manufacturing stage. At this point there is less room for improvement because the cost structure has been determined and only incremental improvements in process are possible.

Figure 3-6 shows the relationship between the three types of VE activities.

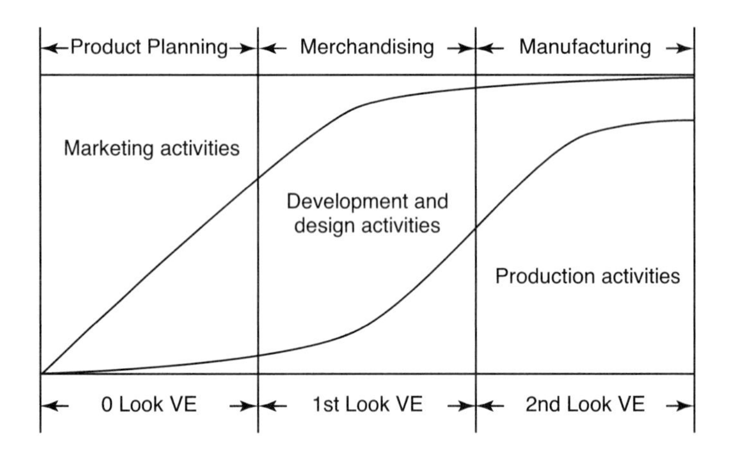

Figure 3-6. Product Life Cycle and VE Activities

It is best to apply VE at the zero-look stage because preproduction is the most efficient point to reduce costs. There is less room for cost reduction later regardless of whether the process is manufacturing, software design, or office practices design. Ishimatsu (1987) lamented that many Japanese companies apply VE downstream. What he identified as second-look VE is often practiced as a part of kaizen costing. While this may be a valid observation, we must also acknowledge that many Japanese companies in assembly-oriented industries practice zero-look and first-look VE in relation to target costing.

Internal and external VE

Here we will refer back to the VE program described in Figure 3-5. VE is conducted both inside and outside the company. Internal VE is conducted both from the top-down and from the bottom-up. The top-down method focuses on various companywide efforts (e.g., improving the rate of capacity utilization, saving energy, using technological information), a long-term program for efficiency improvement (e.g., automation, CIM, or creating unmanned systems), separate VE projects (e.g., product VE, manufacturing VE, distribution VE, clerical VE), and individual cost reduction activities (e.g., reducing cost in the design, manufacture, and purchase of parts and units). In the bottom-up method, individual employees submit VE proposals for action in design, manufacturing, and clerical procedures. External VE has to do with VE practiced by affiliated companies and customers.

VE proposals

VE proposals consist of both idea proposals and action proposals. While improvement activities typically focus on manufacturing, there are other areas with room for cost reduction. For example, over the 1993 fiscal year an electronics company showed the following cash savings from VE improvements: 70 percent in manufacturing, 12 percent in marketing costs, 9 percent in design, 5 percent in clerical activities, and 4 percent in software and other areas. Generally, VE is performed concurrently with one's functionally assigned job, and bonuses are commonly paid for idea proposals.

Organization for Target Costing

Departments usually responsible for target costing are the product design, product planning, and development departments followed by the accounting and production technology departments. To take charge of target costing, a special department is often created. It is refered to as the target costing department, business planning department, or total profit management department.

For target costing to have the greatest success, it is necessary to operate target costing for each product with a matrix organization linking the planning, design, accounting, production, and marketing departments. Independent of target costing, the product manager (PM) is usually responsible for the product from product planning to marketing.

As an example, company Z introduced target costing but could not make it work successfully. A subsequent shift in the organizational structure made the PM, already a powerful manager, responsible for target costing. This single change made the importance of target costing clear to everyone—and the program became more effective. An invoice price system (internal transfer pricing) was also adopted that made the sales manager responsible for sales and the plant manager responsible for production. Accordingly, the production division sells products to the sales division using invoice prices. These two functionally different divisions now compete with each other for profits—and the company gets better results.

Integrating Mid-Range Business Planning into Target Costing

Business performance is strongly affected by changes in both the internal and external environments. For this reason, companies should closely coordinate their mid-range and/or long-range business planning with corporate strategy. During the late 1960s when Japan enjoyed high economic growth, major Japanese companies actively introduced long-range business planning. Peter Drucker contributed greatly to this change through his advocacy of long-range planning.

However, Japanese managers soon realized that long-range business planning was unsuitable in a changing business environment because of the inability to successfully predict very far ahead. Since the 1973 oil crisis, major Japanese companies have emphasized mid-range business planning over long-range planning. In fact, today's typical Japanese companies use mid-range business planning with a long-range vision.

Mid-range business planning gives companies the best venue for working on new product planning—and target costing provides an important mechanism for integrating business strategy to mid-range business planning through the product planning activity. This integration is a primary requirement for the successful implementation of target costing. Indeed, a 1992 survey shows 69.4 percent of Japanese companies integrating mid-range business planning into target costing (Kobayashi et al. 1992).

Mid-range business planning's time horizon is typically three years and includes several types of project planning. One of the most important is the new product development project prepared by the engineering or development department. The business planning department, an upper-management staff department, coordinates the project plans, including the new product development projects, and integrates them into a formal mid-range business plan. The head of the target costing department discusses new products with engineering managers well in advance. This allows the target costing manager to support the project prepared by the engineering department.

The Target Costing Period

The length of the target costing project depends on the product, operation, or project. It generally follows a product's model-change cycle. For example, the target costing period at Kubota (formerly Kubota Steel) is three years (Monden 1989). This rather long period coincides with the time between main model changes. The managers responsible for target costing spend the first year in discussions with the marketing staff to come up with the new model. The second year is spent moving the original idea to the design stage. This is when target costs are assigned and detailed design and costs formulated. Seven to 10 months are then required to prepare for mass production. Very recently, however, the length of target costing projects has generally shortened sharply because of cost and competition.

Case Study: Determining and Analyzing Target Cost

How is the target cost determined within the framework of target costing? How is target cost analysis conducted? To illustrate the process, we will examine a case study from company X, an automaker.

Determining Target Cost

Company X is planning to develop a new automobile—the Crusader. As shown in Figure 3-5, the product development plan is handled by a new product development committee. A project team is formed and a target cost is set. The team consists of several specialists from different departments such as engineering, marketing, purchasing, and accounting.

The product development committee decided to produce an automobile with a planned sales price of ¥4 million. A target profit of 20 percent was applied using the ROS imbedded in corporate strategy. Accordingly, the target profit was ¥800,000. Subtracting the target profit from the planned sales price yielded an allowable cost of ¥3.2 million. Target cost teams were formed and the various team activities began.

Engineers pulled together the drifting cost of ¥3.5 million based on present technological standards. This was ¥300,000 more than the allowable cost. VE projects were conducted several times for cost reduction. The results were presented to the first target costing committee and the plan was studied further to find more room for cost reduction. The cost was reduced by ¥30,000 in upholstery, ¥2,000 in the drive shaft, ¥10,000 in the engine, ¥40,000 in direct conversion costs, and so on. Finally, they were able to reduce costs to ¥3.225 million. These results were presented to the second target costing committee. They could not eliminate another ¥25,000 of costs to reach the allowable cost without changing desired product functions.

The third target costing committee decided that ¥3.2 million would have to be the target cost and that it would be necessary to focus on cost-control activities during the manufacturing process to eliminate the ¥25,000 variance. Thus, ¥3.2 million was set as the target cost for the new model automobile as is shown in Figure 3-7. At the same time, this target cost was considered to be the standard cost—and they placed their hopes on cost control in the manufacturing stage.

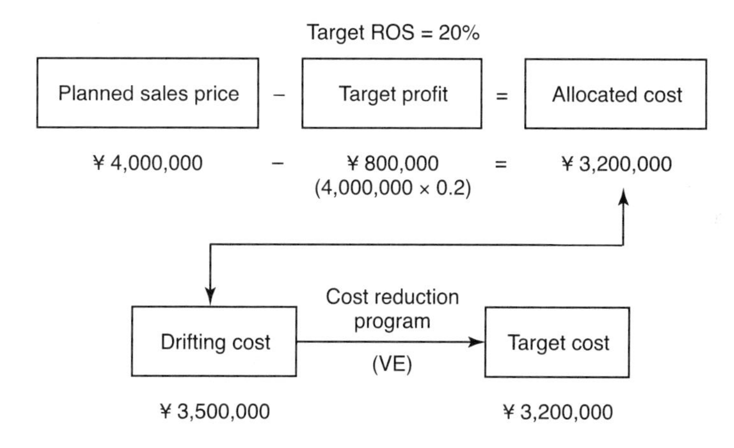

Figure 3-7. The Process of Determining Target Cost

Analyzing Target Cost

As preparations finished, production of the Crusader began. Fortunately, pre-manufacturing VE cost reduction activities were satisfactory; that is, cost reductions to ¥3.24 million were made by this stage.

The new product was very popular. However, production volume dropped because of an economic downturn. Of the original projected sales volume of 2 million cars, only 1.8 million cars were sold. Nevertheless, in light of the economy, the Crusader did well. One month after production began, the actual results were gathered and variance analysis against the target cost was conducted.

The variance analysis led to two findings. First, favorable material and parts costs reduced total material costs by ¥40,000 per car. The result of other cost reduction efforts was ¥10,000 per car. Second, the decrease in sales volume increased the fixed cost per car by ¥60,000. This made the actual result ¥10,000 over target cost per car and made the actual cost closer to ¥3.24 million.

The third target costing committee focused on how to reduce actual cost per car to the ¥3.2 million target cost. As a result, they submitted a proposal to rotate 300 production people to the sales department to handle a sales promotion. Company executives and workers agreed to this project the next month.

Companies Suitable for Target Costing

Is target costing applicable to all companies? Is it suitable for both assembly-oriented and processing-oriented industries? What about software development?

Target Costing in Assembly versus Process Industries

Target costing is effective during model changeovers when there are various products and more opportunities for cost planning to work. Consequently, target costing is often implemented in assembly-oriented industries. In fact, studies from 1984 and 1988 (Sakurai 1984, 1985, 1986; Sakurai and Huang 1988) show target costing applied most noticeably in the automobile, electrical machinery, precision machinery, and electronics industries. The 1992 survey made by Kobe University Researcher's Group supported this finding.

In contrast, mass producing a small variety of products remains the norm in process-oriented industries. It is difficult for target costing to take hold in industries that create added value by processes of this type. The controller at Sumitomo Electric Industries, Japan's largest maker of electric wires and cables, remarked on the difficulty of achieving a target standard at the product design stage in that particular industry (Nagaosa 1980). In short, it is considered difficult to install target costing in process-oriented industries such as steel and chemicals.

However, product diversification in the chemical and steel industries is increasing rapidly. The author's 1988 survey showed target costing installed in 24 percent of the process-oriented companies surveyed, a gradual increase in the application of target costing in this industry. In the chemical industry, 31 percent of responding companies were using target costing in 1992 (Kobe University Researcher's Group 1992).

High Variety/Low Volume Production

Product planning and design is rarely conducted in low variety/high volume production environments. Because target costing begins at the planning and design stages, its importance—and effectiveness—grows in high variety/low volume production.

As the variety of products increases, the need for quality control increases. Likewise TQC is effective in this environment because everyone working together in production contributes to the maintenance of quality. When improvements in quality are undertaken frequently, it is more effective to concentrate on planning than on corrective actions based on feedback information. (Standard costing is the model for this.) In addition, a cost management program that presupposes cost reductions from the very beginning is necessary since technological advances are very rapid. These characteristics all apply to target costing.

Cost Management for Software

Target costing is not a method exclusively for controlling hardware costs. It is also useful for the cost management of computer software. This is primarily because of the importance of cost reduction in the planning and design stages of software development. Since Sakurai advocated target costing's effectiveness (1986, 1987), other papers have been published on this issue.

While conceptually the same as for hardware, target costing for software utilizes different methods. One of the successful applications of target costing is reported by NEC (Matoi 1987). The structure of NEC's target costing is depicted in Figure 3-8.

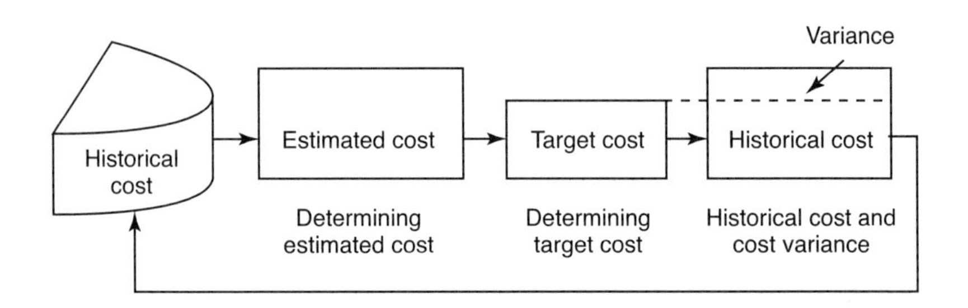

Figure 3-8. Flow of Cost Management for Software

The estimated cost in Figure 3-8 may be considered the drifting cost that was described earlier. Of course, a comparison is made with the allowable cost in proceeding from estimated cost to target cost. Research on cost management for software has been happening for years and many aspects still need improvement.

Conclusion

Target costing is a tool for strategic cost management developed originally in Japan. While it focused primarily on assembly-oriented industries, it is rapidly spreading to process-oriented industries and software manufacturers. Target costing as a process is most effective in the high-tech CIM environment. This does not mean there are no problems with this technique. Regardless of how skillfully a company introduces target costing, it will have no effect unless each and every employee actively tries to reduce costs. Target costing often places unreasonable demands on the workplace. Current research is focusing on overcoming these deficiencies.

References and Further Readings

Hasegawa, Yukio, ed. 1984. *A System Producing a Variety of Products in Low Volume Production Version 2*. The Nikkan Kogyo (June) p. 1.

Horváth, Péter. 1993. *Target Costing*. Stuttgart, Germany: Schäffer-Poeschel, pp. 1-249.

Imai, Kingo. 1987. "A Study of Life-cycle Stages for Each Industry." *Bulletin of the Institute of Business Administration* no. 44, pp. 105-258.

Inaba, Seiemon. 1991. "Target Costing and IE." *IE Review* (May) p. 3.

Ishimatsu, Yasuo. 1987. "Cost Management and Cost Improvement." *The Business and Technology Daily News*. The Nikkan Kogyo, p. 30.

JAA Special Committee. 1994. *Issues to Research in Target Costing*. Japan Accounting Association (June 2) p. 2.

Kan, Yasuto. 1991. "Target Costing Activities in Office Automation." *IE Review* (May) p. 29.

Kato, Yutaka. 1990. "New Developments in Target Costing Activities: The Case of Daihatsu Industries (Ltd.)." *Kaikei (Accounting)* (October) p. 50.

Kato, Yutaka. 1993. *Target Costing; Strategic Cost Management*. Nihon Keizai Shimbun, Inc., pp. 1-328.

Kimura, Katsutoshi. 1992. "Target Costing Activities Around Product Development." *JICPA Journal* no. 440, pp. 61-63.

Kobayashi, Tetsuo, et al. 1992. "A Survey of the Practice of Target Costing." *Accounting* Vol. 44, no. 5, p. 91; Vol. 44, no. 6, pp. 76-77. These surveys indicated the profit planning method (20 percent), engineering method (18 percent), and others (5 percent) in addition to the combination method (57 percent).

Kobe University Researcher's Group. 1992. "Mail Survey on Target Costing (1)." *Accounting*. Vol. 44, no. 5, pp. 88,89.

Kuramochi, Katsuyuki. 1991. "Cost Management in Fuji Xerox's Yuwatsuki Plant." *Journal of Cost Accounting Research*, Japan Cost Accounting Association no. 31, p. 69.

Matoi, Yasuo. 1987. "Software and Target Costing." *Journal of Business Practices* (June) p. 7.

Monden, Yasuhiro. 1989. "Profit Control and Cost Management in Recent Divisional Management Systems: The Case of Kubota Steel." *Accounting* (November) p. 8.

Monden, Yasuhiro. 1991. *Cost Management in Automobile Companies.* Dobunkan Shuppan, p. 22.

Nagaosa, Shigeki. 1980. "Necessary Conditions for Cost Accounting." *Journal of Business Practices* (March) p. 20.

Nishiguchi, Fujio. 1989. "Developing Our Company's Target Costing Program." *Journal of Business Practice* no. 426, pp. 26-29.

Nobbori, Yoshiteru, and Yasuhiro Monden. 1983. "Total Cost Control Systems in the Automobile Industry." *Accounting* (February) p. 106.

Sakurai, Michiharu. 1984. "Cost Accounting for a Variety of Products." *Journal of Cost Accounting Research* no. 278, p. 101.

Sakurai, Michiharu. 1985. "FMS and New Developments in Managerial Accounting." *Accounting* Vol. 37, no. 2, p. 30.

Sakurai, Michiharu. 1986. "How Does FA Change Management Planning and Control Systems?" *Diamond Harvard Business* (Feb/Mar) pp. 67-76.

Sakurai, Michiharu. 1986. "Cost Accounting for Software." *Industrial Accounting* Vol. 46, no. 2, pp. 15-16.

Sakurai, Michiharu. 1987. *Cost Accounting for Software.* Hakuto Shobo Publishing Co., pp. 91-95.

Sakurai, Michiharu, and Philip Y. Huang. 1988. "Cost Accounting in Automated Factories." *The Annual Bulletin of Social Sciences in Senshu University.* Senshu University Social Science Research Institute no. 22, pp. 109-110.

Sakurai, Michiharu. 1991. *The Change of Business Environments and Management Accounting.* Dobunkan Shuppan, pp. 62-66, 314-316.

Seidenschwarz, Werner. 1993. *Target Costing: Marktorientiertes Zielkostenmanagement.* Munich, Germany: Verlag Franz Vahlen.

Survey Office of the Economic Planning Agency. 1982. "Survey of Companies Search Activities, January 1976, and Business Behavior Division at Industrial Policy Bureau in MITI, Company Activity Department." *New Management Indicators*. Ministry of Finance Publication Office (April 11) p. 111.

Tanaka, Masayasu. 1979. "Cost with Target Cost." *Journal of Cost Accounting Research* (October) pp. 37-40, 36.

Tanaka, Masayasu. 1987. "The Evolution of Methods in Cost Engineering in Product Development." *Accounting* (February) pp. 23-25.

Tanaka, Masayasu. 1980. "Cost Control in New Product Development in Japanese Companies." *Accounting* (February) pp. 20-21.

Tanaka, Takao. 1990. "New Product Development and Target Costing in Automobile Manufacturing." *Accounting* Vol. 42, no. 10, pp. 18-19.

The Society of Management Accountants of Canada. 1994. *Implementing Target Costing*. Management Accounting Guideline 28, drafted by Robert A. Howell.

Yamada, Yoshio. 1983. "Our Company's Accounting and Cost Management." *Journal of Cost Accounting Research* (October) p. 12.

CHAPTER 4

Japanese Practices of Overhead Management

In the past 25 years, factory automation, information technology, and globalization have changed the cost and income sructure of Japanese companies by increasing overhead dramatically. As the degree of factory automation increases, so does manufacturing overhead. This increase in overhead along with increased domestic and international competition have decreased the efficiency of operations and management, because overhead is inherently more difficult to manage than direct costs.

Consider the rapid increase of the overhead-to-manufacturing-cost ratio. Manufacturing costs are reported in Japanese annual financial reports as material, labor, and factory overhead. In these reports, factory overhead does not include indirect labor, as is the practice in the United States; instead, indirect labor costs are included in labor cost. In 1965, the overhead-to-manufacturing-cost ratio in all public companies listed on the Tokyo, Osaka, and

Nagoya stock exchanges was only 19 percent. By 1991, the same ratio had gone up to 25 percent, as shown in Figure 4-1 (Tsuji 1993).

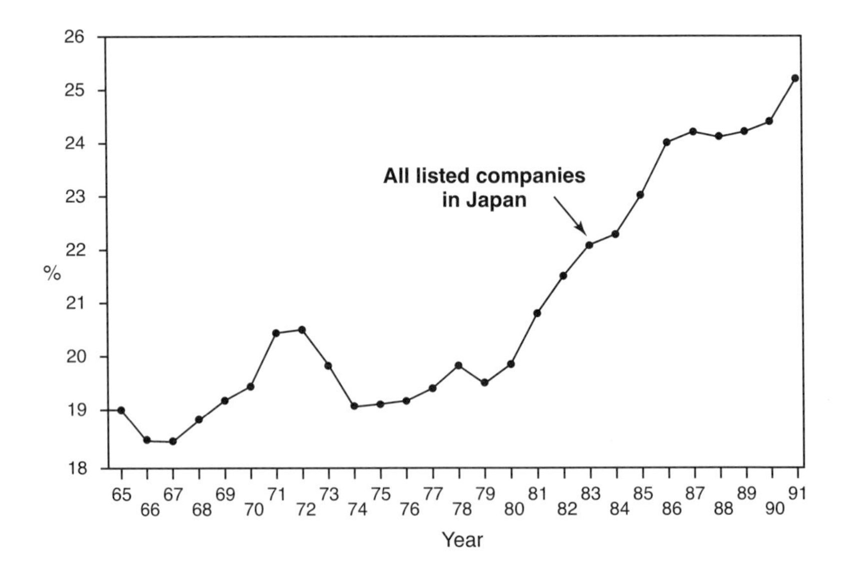

Figure 4-1. Overhead-to-Manufacturing-Cost Ratio

Selling and administrative expenses have also increased in Japanese firms over those 25 years. The ratio of selling and administrative expenses to sales was only 14 percent in 1965. By 1991, it increased to as much as 19 percent. Remarkable increases in overhead cost can be seen in such costs as service-related labor, depreciation, R&D, physical distribution, advertising, and information processing costs. For example, the R&D-to-sales ratio was only 0.4 percent in 1965, but increased to 1.8 percent (a 450 percent increase) by 1991.

As a natural result of this dramatic rise in overhead costs, typical Japanese firms can no longer enjoy the high rates of return characteristic of the 1980s. Indeed, high levels of overhead cost must be one of the most important reasons for the low return in Japanese firms (Ueda 1993).

The purpose of this chapter is to document Japanese practices of overhead management in advanced manufacturing environments.

In August 1989, Robin Cooper and the author visited seven Japanese companies and participated in two academic/practitioner workshops on overhead management. These visits and workshops revealed that many Japanese companies have created innovative practices to manage overhead. In fact, five companies of the seven visited had changed their practices on overhead management within the last three years. Although these changes in cost accounting methods differ, they do have common characteristics. Because they tie cost accounting data with analysis of business activity to enhance cost management, they can be called "activity-oriented accounting."

Overhead Management in Three Japanese Companies

Companies X, Y, and Z share a number of characteristics. Each is the world leader in its industry. Company X is involved primarily in industrial materials manufacturing, company Y is in the machinery industry, and company Z is a typical assembly-oriented industry. We can characterize them as mature companies with very competitive product markets. While each is threatened by competition both domestically and in the newly industrializing countries, the reliability of their products has maintained their competitive edge.

All three companies are highly automated: company X in process automation and factory automation; companies Y and Z in factory automation. Their direct labor has decreased remarkably in the past ten years due to these significant increases in factory automation. Conversely, indirect labor (R&D, planning and design, maintenance, software development, and so on) has increased dramatically. Each company has switched to high variety/

low volume production. These forces increase overhead costs and cause many companies to change their cost allocation systems. Although company X has not, companies Y and Z have changed cost allocation systems within the last three years.

Before the company interviews, a 12-item questionnaire on overhead management practices was administered. Participants were asked about the ratio of overhead to manufacturing costs, their reasons for increasing or decreasing overhead, ways used to manage overhead, allocation methods, their interest in ABC and accurate allocation, the relationship between cost and price, and so on. The results are discussed below.

Company X: Industrial Materials Manufacturing

Faced with rising costs, especially in overhead, company X's management is now very interested in overhead management. Today's overhead is 37 percent of full production cost whereas two years ago it was only 33 percent—a 4 percent increase.

The composition of production costs has changed significantly in five years. Full production cost rose by 9 percent during this period while direct manufacturing costs rose only 3 percent. Two factors drive this small increase, the first being that FA has decreased the number of workers and improved quality. The second factor is that active cost control activities have reduced costs.

Meanwhile, a 22 percent increase in overhead costs in this five year period could be attributed primarily to customer demands for higher quality. For example, the company must maintain certain inventory levels for its just-in-time delivery system (and their JIT suppliers must have enough inventory to supply them). Even dust settling on products during storage is a quality issue—and improving quality means more work in the support and service areas. These activities increase overhead costs.

Company X supplies intermediate raw materials to other manufacturing companies. However, because wholesale product prices are regulated in Japan, it is difficult to compete in the area of price. And, as mentioned earlier, quality standards in the industry are extremely high. Company X's remaining strategy is to provide better customer service. Increases in overhead costs result from the engineering support and services used to improve customer service.

Overhead categories

Company X divides overhead into four categories:

- engineering costs
- production scheduling costs
- employee welfare-related costs
- corporate expenses

Engineering costs cover quality control, maintenance, and material handling. They comprise 27 percent of all overhead costs. Amounts of energy consumed (gas, water, heat) are used as cost drivers.

Production scheduling costs include the engineering and factory planning staff. Production scheduling costs comprise 5 percent of all overhead. Of these, production planning and computer-related costs are attributed to products or product lines by machine-hours. Costs related to factory offices comprise only 1 percent of all overhead costs. It can be difficult and even impossible to find widely supported cost drivers for assigning these costs. Therefore, because they mainly relate to the number of workers, plants, and equipment, these costs are allocated by both labor costs and depreciation.

Employee welfare-related costs include recreation, hospital, housing, and other social benefits.

Corporate expenses are nonmanufacturing costs that include executive salaries, public relations expenses, expenses of the legal and tax department, corporate planning, and R&D costs. Corporate expenses comprise 16 percent of all overhead. They are charged directly to revenue as a period cost for financial reporting purposes. However, they are allocated to products or product lines (using sales amounts and equipment book value as cost drivers) for product profitability analysis and other management purposes.

Managing overhead costs

One of the most effective ways to manage overhead costs is to reduce the number of workers needed to operate the plant. For example, the maintenance department in company X has changed dramatically. Robotization was introduced for repair work and an expert system was applied to the diagnosis and inspection of equipment. Along with these programs, a unique "adaptive maintenance system" is used to manage all maintenance systems. These programs have greatly improved the productivity of maintenance. The company can now systematically measure maintenance activities, including losses from breakdowns and repair costs.

Company X uses a standard cost system. During budgeting, product line costs are computed by department. However, historical cost is computed only by cost elements and by department—there is no monthly closing for product line reporting. Inventory valuation is based solely on standard cost. System costs are divided into variable and fixed costs and the contribution margin is used for managerial purposes. The purpose of variable costing is to evaluate short-run production levels and to determine the best production capacity, staffing level, and product mix.

With selling prices determined by official quotation, cost has no role in pricing decisions. In fact, cost plays an important role in only two cases. The first is when new products are developed. In this instance, production cost can be a part of the basic data used by plant and department managers to determine the selling price. The second case is when profitability of product lines is measured. To maximize companywide profit, the company introduced a "profitability control system by product lines" that displays income by product lines in addition to divisional income (*The Nikkei Financial Daily* 1989). This new system allows company X to immediately cope with new market developments in a strategic manner.

Company Y: Machinery Industry

Although company Y effectively manages variable costs with target costing, it lacks an effective technique for managing overhead. Increasing pressure to reduce overhead due to the adverse exchange rate and domestic competition now makes overhead management a significant managerial issue.

Overhead constitutes almost 25 percent of full production costs, which seems lower than the Japanese average. However, overhead in the machining industry runs about 25 to 30 percent of full production costs. In recent years, the proportion of overhead to production cost has gradually increased.

Long, mid-, and short-term effects of either increasing or decreasing overhead costs must be identified. From a long-term perspective, introducing FA increases depreciation and indirect labor while decreasing direct labor. The increased number of machines and robots raises their ensuing administrative costs. At the same time, reduced expenditures for items like imported supplementary material, heavy oil, welding sticks, and power can be enjoyed. A 1994 survey of manufacturing technology in *The Economist* (1994) notes that because robots are more consistent than humans, designers can reduce the number of spot welds in a design by as much as 10 percent—as well as associated direct and indirect costs.

From a mid-term perspective, the increased volume of low-priced imported goods does not reduce administrative costs. This means that administrative overhead cannot be reduced proportionally with decreased import prices.

In the short term, the labor cost for model and design changes increases. As product variety increases in response to the market, non-value-added activities are likely to occur because of the frequent order changes and changes in production plans. Summing up, total overhead increases slightly.

Managing overhead costs

Company Y basically manages overhead in each cost center using budgeting. While they do perform a post-audit of investment in plant and equipment, no strict performance evaluation or follow-up is performed. In general, department managers are given a free hand in spending—and unless there is an overbudget expenditure, there is no review. Overhead is managed by everyday operational activities using physical data, not financial data. Even without converting overhead cost reduction into dollar amounts, departmental costs and subsequently unit production costs can be reduced if everyday cost reduction activities are effective.

Company Y's machine centers are cost centers. Contrary to usual Japanese practice, ROI is emphasized—although it is not used on the plant floor to measure performance. This is because the company produces multiple product lines from the same plants, making it difficult to separate the investment in any one line.

As part of their search for effective overhead allocation, company Y's managers examined all then-available information on activity-based costing. However, they concluded that they did not need "accurate" and "sophisticated" assignment. Instead, they needed an "appropriate" and "simple" method. By this they meant that they only needed to know which cost in a department related to a product line—not to a product item. They also defined separate cost centers only when overhead by product lines was significantly different and identifiable. To keep things simple, they also decided to select direct charge or simple allocation bases (because complex allocation bases were hard for floor managers to understand and interpret and also hard to collect and manage).

Company Y changed its allocation method in 1989 to include some heavily modified ABC principles. The company no longer uses a two-stage allocation system. Instead, it assigns indirect costs to the product line using appropriate cost drivers. For example, engineering work is divided into three categories:

engineering work, service work, and factory administrative work. Engineering work costs are charged directly to product lines. Service work costs are attributed using rates based on standard labor-hours, machine-hours, and weight as needed. Factory administrative costs are allocated to product lines using conversion costs as an allocation basis.

The new system is unique in three ways. First, the company has discontinued its two-stage cost allocation system. Second, it uses conversion cost as a cost classification instead of direct labor and overhead. This saves administrative effort and is becoming more common as FA reduces the amount of direct labor. Third, rather than assigning costs to products, it assigns them only to the product line ("direct charge of overhead to product line system" or DCOPLS).

These changes are not unique to company Y. Typical Japanese companies are currently discontinuing the two-stage allocation system and implementing the concept of direct charge to product line. Most Japanese managers now consider it best to connect costs to business activities for cost control. The basic mechanism for DCOPLS is depicted in Figure 4-2.

Company Z: Assembly-Oriented Industry

Overhead management is likewise important in company Z. However, the issue of overhead allocation has been insignificant for several years—overhead as they define it is only 4.5 to 5 percent of full production cost. This level is low because they directly charge all costs incurred in a direct production department to the product line using DCOPLS (similar to company Y). For example, the depreciation cost of machines constitutes 3.5 percent of full production cost and is charged directly to the product line as a direct cost. Tools are also treated as direct costs and their costs excluded from overhead. This accounting treatment is not unique to companies Y and Z—many Japanese companies follow it.

Naturally, changing the cost object from the product to the product line has changed the definition of "direct." Let's look at

Old System

Engineering work
Allocated engineering costs
Stage 2 (allocation)
Product
Service work
Stage 1 (allocation of overhead)
Administrative cost
(allocation to product)

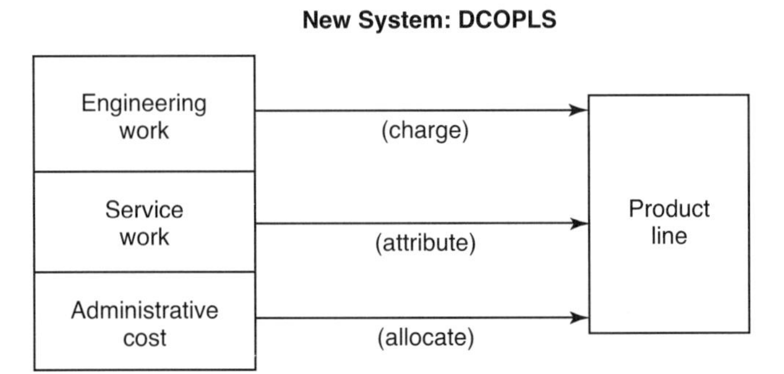

Figure 4-2. Engineering Work Cost Assignment

"direct labor" as an example. Maintenance or material-handling workers are direct labor with respect to the product line even though they are indirect with respect to each product. As a result, company Z's overhead consists of only small amounts of indirect labor and other miscellaneous expense, about 5 percent of full production cost.

However, even this small amount of overhead has decreased about 20 percent in the last five years due to two primary factors. First, the company has improved physical distribution. Its nearly zero inventory has made warehousing unnecessary, allowing them to eliminate overhead costs related to these activities. The second factor is that line workers are trained to be multifunctional and so do both service and support work. For example, because the line workers can operate the robots, the need for special maintenance workers is eliminated. This also applies to quality control; most of the quality work is performed by these same line workers. The costs of sample testing and other activities are charged directly to the product line.

Since the company started using DCOPLS, the cost accounting structure has changed. Even accounting for direct labor has changed, although it is conceptually the same as in other companies. Since direct labor now includes maintenance and material handling, direct labor cost cannot be computed as before by multiplying direct labor-hours by the labor rate. As a result, company Z's management is abolishing the work measurement for standard costing—thereby changing the standard costing structure.

Without allocation in the traditional sense, the company has abolished the typical overhead allocation procedure that was conducted about ten years ago. In other words, the company does not use labor rate or machine rate as an allocation basis. They divide the remaining overhead into several categories and allocate the costs by the ratio of direct costs (to be discussed later). The management team has no interest in activity-based accounting (ABC), feeling that DCOPLS makes the whole issue moot. While it is true that many allocations are necessary when allocating overhead by product, traditional-style allocation is not needed if overhead is attributed only to product lines. Furthermore, because overhead management is conducted fairly well (primarily through budgetary control), overhead allocation is not management's major concern.

Surveys show that the joint use of labor-hours and machine-hours for allocation purposes has become popular among Japanese companies (Sakurai & Huang 1988; Sakurai 1992). However, company Z's management shows no interest in this method for two reasons. The first is because they treat many overheads as direct costs to product lines, resulting in indirect costs that are small in number and amount. The second reason is that even if you try to allocate as accurately as possible allocation's arbitrary nature is inescapable—even with ABC.

Target costing is the indispensable technique for controlling variable costs at company Z. It results in cost-effective technological innovation—unlike accurate allocation, which contributes to neither technological innovation nor cost control. While accurate information in other companies may help to measure product profitability, company Z's management does not make sales-mix decisions using cost data alone. This is because cost in their industry has no direct relationship with selling price.

The concept of contribution margin is important at company Z where they have developed a unique application of it. Their product line contribution margin is the residual of sales amounts minus the direct material, direct labor, depreciation, and applied R&D cost of the product line. Those costs deducted directly from sales are the object of target costing. The contribution margin minus overhead costs is called the operating profit in company Z. The most important concept of profit for the plant managers is contribution margin by product line.

As noted, some allocation amounts are small. For example, plant manager salaries are factory overhead and therefore allocated to product lines using direct material cost as a cost driver. Many Japanese companies have begun to use direct manufacturing cost as a cost driver for the allocation process despite its obvious weakness. If the amount to be allocated is small, and/or the company does not base pricing decisions on cost data alone, then inaccuracies do not constitute a big managerial headache. However, this is not to say that the company does not take cost data into account when making pricing decisions. Rather, it means that cost data only plays a small role in product pricing in a competitive market.

Characteristics of Japanese Overhead Management

We have identified three main characteristics of Japanese overhead management. They are: (1) cost reduction oriented cost accounting systems, (2) product line costing systems (DCOPLS), and (3) less reliance on accounting figures for certain decisions. These characteristics are discussed next.

Cost Reduction Oriented Accounting Systems

In a U.S. survey, over half of the respondents were dissatisfied with their product costing and wanted to develop alternative bases for assigning overhead costs (Howell, Brown, Soucy and Seed 1987). In contrast, their Japanese counterparts did not seek accurate cost data. There was a significant difference between American and Japanese management attitudes toward "minimizing the allocation of indirect costs" (NAA Tokyo Chapter 1988). Survey results indicated that Japanese managers favored minimizing allocations of indirect cost much more than American managers. Even in our visit to seven major Japanese companies, only one manager (in a government-regulated company) showed interest in a more accurate assignment method. It may be because Japanese companies have fewer managers and accountants than American companies (Dixon, Nanni, and Vollman 1989) or because Japanese managers believe that allocation inevitably is arbitrary even when done accurately.

Most Japanese managers want to use cost data for cost reduction—not product costing. The reason why almost all the companies we visited try to attribute costs to product lines is because they want to manage overhead as well as possible. Even though the high variety/low volume production revolution is an underlying factor, we should not forget that the aim is to manage cost rather than to have accurate cost data for profitability analysis.

Why did American managers want more accurate product cost data in the late 1980s than Japanese managers? We suspect they wanted it for better resource allocation or product profitability analysis. In general, U.S. companies suffered in the late 1980s and this may be one reason why "restructuring" was so popular in that period. ABC was an effective tool for this purpose.

During this same time, Japanese companies prospered. Some Japanese companies needed accurate data for sales-mix decisions.

Company X is an example. There we observed the pressure from corporate management to develop profitability analysis reports for restructuring their products. We could have predicted this because company X, competing in a declining market, is under great pressure to determine new sources of profit. However, this was not typical in Japan and may be why most Japanese managers showed little interest in ABC.

When the business environment is unhealthy, companies may be pressured by stockholders to be as profitable as possible. However, Japanese managers typically base their primary strategies on long-term corporate existence rather than product profitability. Because of the inherent emphasis on long-term coexistence with customers, workers, bankers, and investors, they will often continue production even when a product line's profitability is relatively low.

Product Line Costing System (DCOPLS)

Quite a few Japanese companies have switched cost objectives from the product to the product line. An underlying reason is that high-variety production makes computing product costs extraordinarily time consuming. Also, since the charge is directly to a cost objective, from a cost management perspective it is easier to reduce cost through DCOPLS because the relationship between expenditure and activity is more easily identified.

This chapter has discussed how three companies make their product line assignments. They each assign costs to product lines: Company X analyzes profitability by product line; company Y divides factory administrative costs into product lines using direct manufacturing cost as the cost driver; and company Z treats all line costs as direct costs to the product line. Company Z even treats plant and machine depreciation as direct costs and therefore does the least allocation possible. Companies Y and Z use direct manufacturing cost as a cost driver primarily because the overhead amounts to be allocated are minimal.

To show how costs are charged directly to a product line, we constructed the model for product line cost accounting in Figure 4-3.

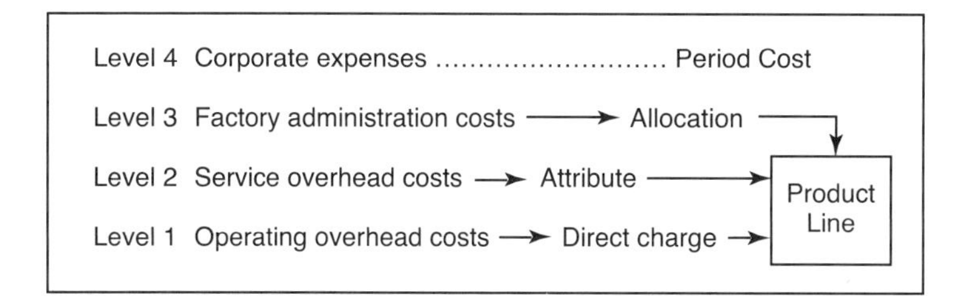

Figure 4-3. Structured Model of Overhead

As illustrated, costs from level 1 to level 3 constitute manufacturing overhead. Typical costs for level 1 are setup cost and maintenance cost. Because they are directly used by a product line, they can be charged directly to each line. Typical costs for level 2 are production scheduling and inventory control. These cost are attributed to a product line by finding appropriate cost drivers.

Typical costs for level 3 are salaries for plant managers and cost accountants. Because of the difficulty in attributing these costs directly to a product line, they must be allocated. Level 4 costs include corporate expenses, basic R&D, and indirect sales promotion costs (such as advertising and entertainment). However, these costs are treated typically as period costs and must be allocated to a product line when information is needed for profitability analysis. For example, advertising cost for a specific product line can be treated as a direct cost—but not the cost of image advertising. The same is true of basic R&D and corporate administrative costs, which must be allocated by arbitrary allocation methods.

Less Reliance on Accounting Figures

Both American and Japanese academics and practitioners believe that cost is not the only factor for determining price. However, more American accountants seem to think that accounting figures play an important role in pricing decisions and product profitability analysis. They feel that management can set prices and form a strategy based on cost information. In fact, much of the discussion about the effective use of ABC relies implicitly on this assumption.

On the other hand, almost every Japanese manager we visited denied the relationship between cost and price except in the development of new product categories and product profitablility analysis. They do not rely on cost data as American managers do (NAA Tokyo Chapter 1988). Instead, they try to reduce overhead on the assumption that the price is a given.

Conclusion

In examining three Japanese practices of overhead management and their characteristics, we found managers more concerned with cost reduction than with product mix decisions—largely as a result of Japan's healthy business environment in the late 1980s. Japanese managers are strongly oriented toward activity types of analysis because cost accounting's primary goal historically has been cost control or cost reduction.

The easiest way to control cost is by knowing the cost of achieving cost objectives. High variety/low volume production has made it difficult for managers to assign costs to cost objectives. For this reason, the three companies studied in this chapter assign costs to product lines—not to products.

When managers rely heavily on full cost data for profitability analysis or pricing decisions, they must have accurate information on product cost. However, Japanese managers placed less importance on accounting data for these types of management decisions. Instead, they valued nonfinancial data or physical, first-hand data. Moreover, in cases where they used cost data for profitability analysis or pricing, they used variable cost or contribution margin data.

Japanese companies are now struggling to recover from the deep economic depression of late 1991. They needed to restructure their product mix and revitalize a bad business environment. As a result of this change, some Japanese managers began to look at activity-based costing. This will be discussed in the next chapter.

References and Further Readings

Dixon, J.R., A.J. Nanni, and T.E. Vollman. 1989. "Breaking The Barriers: Measuring Performance for World Class Operation." Working Paper, Graduate School of Business, Boston University.

Howell, Robert A., James D. Brown, Stephen R. Soucy, and Allen H. Seed III. 1987. *Management Accounting in the New Manufacturing Environment.* A joint study by the National Accounting Association and CAM-I.

NAA Tokyo Chapter. 1988. *Management Accounting in the Advanced Manufacturing Surrounding: A Comparative Study of a Survey in Japan and the United States.* IMA Tokyo Chapter (October) pp. 14-15.

(1) The following are comparative results on allocation measured by number of companies: extremely important (13 U.S., 18 Japan); very important (21 U.S., 10 Japan); somewhat important (15 U.S., 8 Japan); not important (13 U.S., 0 Japan).

(2) The following are comparative results on cost data: very satisfied (12 U.S., 0 Japan); seems reasonable (31 U.S., 50 Japan); needs improvement (47 U.S., 48 Japan): dissatisfied (7 U.S., 0 Japan).

The Nikkei Financial Daily. 1989. "Profit Maximization by Product Line: Market-oriented Approach." (February 7).

Sakurai, Michiharu. 1989. "Target Costing and How to Use It." *Journal of Cost Management* (summer) pp. 39-50.

Sakurai, Michiharu and Philip Y. Huang. 1988. "Practices of Management Accounting in a FA Plant." *Business Review of Senshu University* (September) p. 63.

Sakurai, Michiharu. 1992 "Japanese Management Accounting Practices: Analysis of CIM Mail Survey." *Business Review of Senshu University* (October) pp. 118-119.

"A Survey of Manufacturing Technology." 1994. London: *The Economist* (March 5).

Tsuji, Masao. 1993. "Structural Change in Cost and Revenue of Japanese Companies." In "A Special Committee Report of

Japan Accounting Association." *How Cost Management Systems Should be Structured under the New Business Environment.* Chaired by Michiharu Sakurai, p. 85.

Ueda, Nobuhiko. 1993. "Restructuring of Business Management and Its Future Outlook." *JICPA Journal* no. 56 (June) p. 10.

CHAPTER 5

Activity-Based Costing and Overhead Management

The biggest issue facing business today is the size of overhead consumption and the lack of effective tools for managing overhead costs. As discussed in earlier chapters, direct manufacturing costs can be reduced effectively using target costing and/or standard costing. Both techniques are pervasive in Japanese companies. However, it is difficult to reduce the growing size of overhead by using these techniques since their primary focus is on materials, parts, and direct labor cost. Although many Japanese accountants use budgets for managing factory overhead, budgets themselves do not serve other cost accounting goals (such as assigning appropriate costs to products) nor are they an effective tool for cost analysis. In addition, most traditional volume-driven cost accounting systems do not appropriately assign factory overhead to a product for cost analysis. This is especially true in today's diversified manufacturing environment.

Many approaches have been proposed to cope with overhead measurement and management within the new manufacturing environment. Activity-based costing (ABC) is the U.S. method that has attracted the most attention recently as a solution to the measurement issues. Aimed at process management, activity-based management (ABM) has evolved naturally from ABC. In the course of this chapter we will examine the different response to ABC/ABM in the United States and Japan.

The different receptions to ABC/ABM seem based on three related themes that appear several places in this book. The first is the strong Japanese orientation toward letting the cost management impulse create demand for appropriate cost measurement. This is in contrast to the U.S. tendency to assume that precision in cost measurement will motivate management effectiveness. The second theme is the cost reduction focus of Japanese managers that, in this context, reduces the emphasis placed on product mix and pricing decisions. The mission of organizational survival in the Japanese sense (described in Chapter 1) leads Japanese managers to focus on continuous cost reduction activities rather than on short-term product mix decisions. Thirdly, when ABC first appeared, most Japanese companies were enjoying a healthy business climate and saw little to interest them in the incremental benefits ABC offered for restructuring business organizations.

However, with the dramatic changes that followed the 1991 collapse of the bubble economy, many Japanese firms now struggle for survival. The stark new reality has made the additional benefits offered by ABC/ABM more attractive than ever (even though the Japanese approach to information management means that ABC analysis usually leads not to a transaction-based accounting system but to specific cost reduction decisions in the ABM fashion).

Three Approaches to Overhead Management

Quite a few managers in manufacturing companies in the United States (Howell et al. 1987) and Japan (NAA Tokyo Chapter 1988) are dissatisfied with their current cost accounting systems. What disturbs them most is the allocation and management of factory overhead. Three approaches are considered possible remedies to these problems: variable costing, improved allocation procedures, and ABC.

Variable Costing

The first and most obvious solution is to discontinue allocation entirely—since it may be "the root of all evil." This approach ultimately leads to the adoption of variable costing (direct costing). During those periods in the past when variable costs were proportionally larger than fixed manufacturing costs, variable costing information was appropriate because most costs were variable and therefore controllable in such a system. However, in the FA environment, the proportion of fixed costs is extremely high and variable cost information by itself is no longer as useful for cost analysis (Tokai 1983; Dilts and Russel 1985; Sakurai 1986; Kaplan 1990a.)

Also variable costing is considered to be most effective for short-term decisions in later stages of a product's life cycle (maturity and decline). Yet the modern marketplace shows that many products have extremely short maturity and decline phases, reducing the impact of variable costing. In addition (and in contrast to traditional economics), many key managers consider it both useful and safe to allocate factory overhead to products in order to estimate long-term average costs. This means that variable costing cannot provide a comprehensive solution to the overhead accounting issues—even though it does give useful information for making pricing decisions and cost analyses, particularly in certain process-oriented industries.

Improved Overhead Allocation

A second possible remedy is to improve the allocation method. While there are several ways to do this, the most popular are the use of machine-hours in addition to labor-hours and direct assignment to product line.

Machine-hours in addition to labor-hours

This is a partial improvement of the existing allocation method. It directly responds to the mechanization in the new manufacturing environment by using machine-hours or both labor- and machine-hours rather than just labor-hours as the allocation base. The use of machine-hours is a traditional response to mechanization that was observed in America as early as 1901 by Hamilton Church (1901) and in 1906 by John Whitmore (1906). However, it remains a volume-based method and does not fully address the problems inherent in modern high-overhead environments. If engaging in high variety/low volume production, these factories use large amounts of non-volume-related costs. A capital intensive operation engaged in mass producing a small variety of products probably would not experience the same level of non-volume-related costs—making it a candidate for this method.

Direct assignment to product line

As discussed in Chapter 4, this second improvement approach minimizes the allocation of factory overhead and makes assignments to a product line rather than to individual products. This method is allowed by the Cost Accounting Standards established in 1962 by the Business Accounting Council of Japan's Finance Ministry (Hirabayashi 1989).

The setup costs frequently discussed in this chapter must be "allocated" if individual products bear the full cost. On the other hand, if assigned only by product line, these cost need not be allocated. (Companies X, Y, and Z in the previous chapter used this method.)

This method is employed in major U.S. companies as well as in Japan. For instance, an American IBM keyboard plant has switched from traditional allocation methods to directly charging each product line for its maintenance and quality control costs. As a result, 75 percent of factory overhead is now charged directly to a product line as compared to previously being allocated using arbitrary methods (Drury 1989). Similarly, a method of direct charging is used at Hewlett-Packard (Hunt, Garrett and Merz 1985). Quite a few Japanese practitioners regard this as one of the most promising ways to manage overhead in focused-factory organizations.

Activity-Based Costing (ABC)

A third approach to managing overhead is to fundamentally revise the method and philosophy of assigning overhead. The activity-based costing (ABC) approach proposed by Robin Cooper and Robert S. Kaplan is such a method. Originally practiced by Hewlett-Packard, John Deere, Siemens, GM, and other major U.S. firms, ABC is probably the most frequently discussed innovation in U.S. management accounting in the 1990s. A variant of ABC called *Prozeßkostenrechnung* (process cost accounting) was developed in Germany by Péter Horváth and Reinhold Mayer (1989).

Broadly speaking, there are three categories or approaches to ABC:

1. *activity accounting* proposed by CAM-I (Berliner and Brimson 1988) and James A. Brimson (1991)
2. the original *activity-based costing* (ABC) advocated by Cooper, Kaplan, and followers since the late 1980s
3. *activity-based management* (ABM) proposed by Kaplan, Peter Turney, and others mainly since 1991 (also called activity-based cost management or ABCM)

Aside from the challenges inherent in any cross-cultural comparison, the issue is further clouded because several slightly different versions of ABC have evolved over a period of years. For example, even the basic characteristics of ABC have been described differently over time. What were originally presented as allocations are now called estimates (Kaplan 1992) and the term "allocations" is forcefully proscribed. Acknowledging the diversity of views and the rate of change in writing about ABC, this discussion will focus on a generic description of ABC and then skip to a discussion of ABM (which has been received quite differently in Japan than ABC).

While the most influential new technique in the late 1980s was ABC in its original form, by the early 1990s ABM is more influential. Pinning down the relationship between ABC and ABM is difficult because several proponents claim ownership of and definition rights over these terms—and they do not always agree. However, there are three main views about the relationship between ABC and ABM in the United States and Japan. One is the argument raised by CAM-I (Raffish and Turney 1991) that defines ABC as "a subset of ABM" even though ABM followed ABC historically. A second view, held primarily by ABC proponents, argues that ABC is necessary for ABM and is how ABC provides value. (In other words, ABC and ABM are most effective only when they are tightly linked.) Some regard them as complementary—ABM cannot be installed effectively without ABC, and ABC alone is not beneficial to an organization. Others view ABM as evolving from ABC without being necessarily tied to it.

This last view differs from that held by proponents of ABC/ABM/ABCM. Based on discussions with academic colleagues and business contacts, it represents the views of many knowledgeable Japanese and U.S. accountants who do not completely agree with the core writing in this area.

The ABC originally described in seminal articles emphasized how measurement-based management (better overhead allocation) could lead to increased profitability, usually through product mix and pricing decisions as well as cost reduction efforts. This may be one of the major reasons why Japanese managers reacted negatively to ABC proponents such as Cooper in the late 1980s. Clearly, the *gee whiz*! impact of ABC on price and product mix heavily influenced readers' understanding of it. Kaplan (1992) acknowledges that in their original work, he and Robin Cooper focused on product costing before they realized the equal importance of a better understanding of activity and process costs.

In any case, current views see ABC as an information system and ABM as a set of practices or actions based on activity-based concepts. Some proponents claim that ABM is exclusively and totally bound to ABC as its information source. Others argue that only ABC concepts and some ad hoc analyses are needed for effective ABM. However, we believe that while ABM evolved out of ABC, their major thrusts differ. ABC's major purpose is to provide managers with product cost information for product profitability analysis and other decisions. In comparison, ABM aims at cost management for process improvement and innovation.

Original ABC

Producing a wide variety of products in small lots is typical of the new manufacturing environment. Support activities have also increased (such as setup, planning and design, engineering, and material handling). As a result, two questions arise:

1. Are traditional methods of assigning factory overhead for measurement purposes satisfactory in today's environment? (ABC focus)
2. Is the approach to managing overhead satisfactory with existing methods? (ABM focus)

Reasons for the Original ABC

Traditionally, factory overhead is allocated using a departmental application rate to determine the amount allocated. This approach has two stages. In stage 1, costs are assigned to a cost center. In stage 2, these costs are allocated to products. Figure 5-1 illustrates this process (Cooper and Kaplan 1988).

In stage 2 of the traditional process, the allocation basis used most frequently is direct labor-hours (DLHs). While this was appropriate when much of the value was added by direct labor, today's manufacturing environment shows the number of workers decreasing and that of industrial robots increasing. As a result, the proportion of direct labor cost to total manufacturing cost has dropped significantly in contrast to a growing proportion of indirect support function costs.

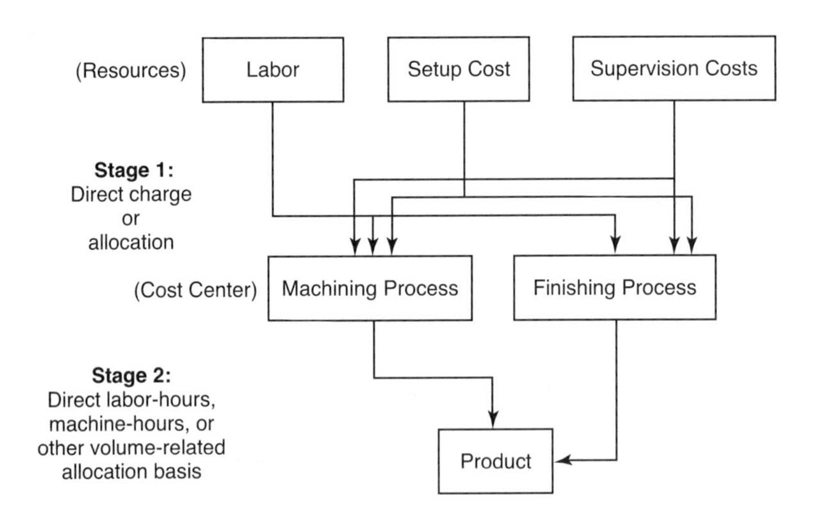

Figure 5-1. Traditional Cost Allocation

Productivity in a FA or CIM manufacturing environment is no longer based primarily on employee efficiency. Instead, it is determined by a number of factors—factors such as the performance of the machine itself, the quality of the software, the quality of the employees, and the quality of maintenance. Since, in most cases, Japanese shop floor employees successfully operate the machines, the skill level of the computer experts and engineers has the greatest influence on productivity.

In today's machine-focused manufacturing environment, we can no longer assume that allocating manufacturing costs by direct labor-hours is rational in every instance. A British-U.S. study (Howell, Brown, Soucy, and Seed III 1987) indicates a tendency toward the increased use of an allocation method based on machine-hours. In contrast, very few Japanese companies use only machine-hours. Rather the Japanese trend is to jointly use direct labor-hours and machine-hours (Sakurai 1986; Sakurai and Itoh 1989).

In traditional volume-driven cost accounting systems, the accounting process focuses on products rather than on activity. Since it was presumed that products would consume prime costs, cost followed products. Thus, the traditional allocation basis was a volume-related factor such as direct labor-hours, machine-hours, or materials consumed. Cooper and Kaplan emphasize the need to move away from this traditional method of allocating overhead in situations where volume does not drive cost.

In ABC, the activity itself becomes the focus of the costing process. Costs are traced from an activity to a product by determining how much of the activity the product requires. Thus, the allocation basis used in ABC is a measure of the activity performed. Among the types of activities mentioned in ABC discussions are setup time, how often orders are placed, number of parts, and processing time. All overhead costs are assigned first to the major manufacturing processes (or "activity centers" in ABC). Activity center costs are not usually consumed proportionally to the number of units produced because different cost objects trigger resource consumption. In stage 2 of ABC's allocation process, the costs of every different activity performed by each center are assigned to products using the number of transactions required to perform every activity for each product. These second-stage drivers can be setup time, warehouse moves, inspections, or sales calls—along with direct labor-hours or machine-hours. Figure 5-2 shows ABC's cost assignment process.

Basic Concepts of the Original ABC

The basic elements of ABC are the cost driver and resource consumption. Our examination of these concepts is based on an interpretation of the original ABC model proposed by Robin Cooper and Robert S. Kaplan.

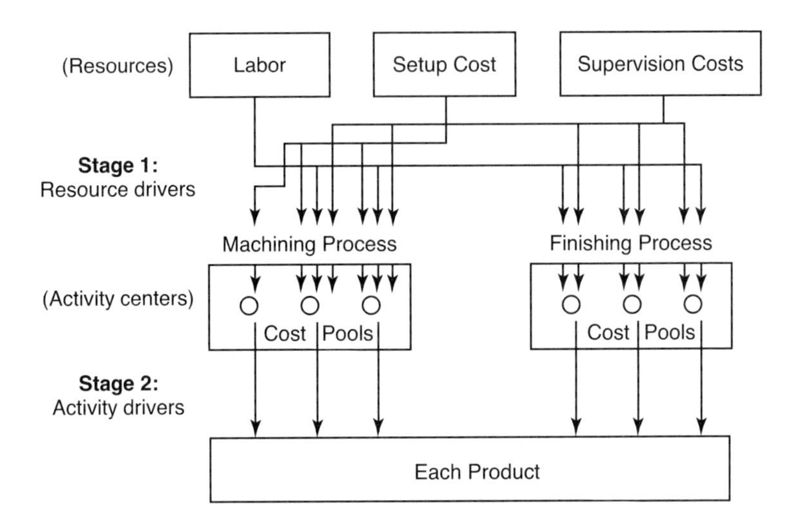

Figure 5-2. Cost Assignment in ABC

The cost driver

A cost driver is any factor that causes a change in the cost of an activity. Cost drivers are the causal events that influence the quantity of work (and therefore costs) in an activity (Raffish and Turney 1991). The term is applied to two situations. A "resource driver" deals with the accounting mechanism of assigning resources to activities. The "activity driver" deals with the accounting mechanism of assigning activity costs to cost objects in the system.

Cost drivers are of two types—volume-related and non-volume-related. An example of a volume-related cost driver is that materials are consumed in direct proportion to the amount of production. The non-volume-related drivers are a heterogeneous lot. Examples are the cost of the support functions of design, testing, materials handling, and production planning. While the cost of these activities does not normally change in the short run, it does increase as the production process becomes more complex and as more demands are placed upon the support functions.

They are traditionally considered to be fixed costs or step fixed costs—although they exhibit a wide variety of behaviors. It is in describing and understanding the variety and subtlety of the non-volume-related cost drivers that ABC literature presents its most sophisticated contributions to accounting.

Another significant difference between traditional costing and ABC is the concept that costs are hierarchical. In ABC, activities are categorized into a taxonomy—unit, batch, product, and facility-sustaining—for the activity cost drivers used in activity-based cost systems (Cooper 1990). Customer-driven activities are divided into four categories—order level, customer level, market level, and enterprise level (O'guin 1991).

Resource consumption versus spending

ABC is also differentiated from its product-centered counterparts by being a model of organizational resource consumption —not spending. For example, a production machine consumes energy as it processes parts. When more parts are processed, more energy is consumed. Of course, spending is ultimately tied to consumption. However, understanding the distinction is very important for decision making because (1) the time lags between consumption and spending can be significant, and (2) consumption and spending can occur in separate cost centers, which means that consumption is never correctly associated with spending. A focus on consumption allows the development of a much more accurate model and reveals why operational improvements often do not lead to lower spending.

Japan's mandatory Cost Accounting Standard defines cost as the monetary expression of goods and services consumed for *Leistung* (cost objectives) in business organizations (Cost Accounting Standards, 1.1.3., 1962). Since all Japanese public companies must follow this standard closely, the present Japanese cost accounting systems already separate consumption from spending. In other words, resource consumption and resource spending in theory and practice are clearly divided in Japan—even without

ABC. All Japanese companies are supposed to measure the costs of resource consumption by individual products or product lines and not by resource spending. One implication of this characteristic is that, as Japanese managers repeatedly and explicitly state, ABC in Japan does not provide the dramatic improvement experienced in the United States.

Practically, however, ABC could improve Japanese cost systems. Current systems do not separate unused capacity from resources used, for example. A major theoretical advance in ABC is that the ABC system should not assign all organizational expenses to cost objectives. Thus, the activity-based system can measure the costs of using resources, not the cost of supplying resources. These two quite different cost concepts are linked via the following simple and fundamental equation (Kaplan 1994):

$$\begin{matrix}\text{Cost of} \\ \text{resources} \\ \text{supplied}\end{matrix} = \begin{matrix}\text{cost of} \\ \text{resources} \\ \text{used}\end{matrix} + \begin{matrix}\text{cost of} \\ \text{unused} \\ \text{capacity}\end{matrix}$$

Unused capacity is a sort of non-value-added activity (capacity). If the unused capacity concept is applied to non-value-added analysis, ABC can be used to eliminate waste. In fact, measuring unused capacity may be one of ABC's most important contributions to Japanese companies. For example, Sanyo Electric, a major consumer electriconics maker, installed ABC/ABM in July 1994 and successfully used ABC to find non-value capacity.

Three Different Examples of Overhead Allocation

The original ABC differs most from the traditional method in its definition of the cost activity centers and in stage 2 of its cost assignment method. To make this clear, the following sections present examples of second-stage allocation based on (1) the traditional method (direct labor-hours), (2) the modified traditional method (both direct labor-hours and machine-hours), and (3) the

ABC approach. A simplified example is used to introduce the essence of these methods. Table 5-1 shows data for production volume, activity centers, allocation bases, and overhead.

Table 5-1. Production Volume, Activity Centers, Allocation Bases, and Overhead

Pro.	Prod. V.	Assembly DLH	Machine MH	Design Number	Setup Number	Purchasing Orders	Mtl. hdlg. No. of parts	Total
P_1	10	1	1	1	1	1	3	
P_2	100	9	10	1	2	3	7	
Total	110	10	11	2	3	4	10	
Cost		$800	$550	$420	$270	$240	$520	$2,800

The Traditional (Direct Labor-Hour) Method

Most American and Japanese companies traditionally have used a volume-related overhead allocation framework based primarily on labor-hours. In stage 1 of the allocation process, overhead is assigned to the production departments. (For example, rent is allocated by space, human resource expenses by number of employees, depreciation of machinery and insurance costs according to the value of each work group's machinery, and so on.) Once factory overhead is allocated to the production departments, the second step of the allocation process apportions the department's total cost to each product using an allocation basis related to production volumes. The allocation basis commonly used is the direct labor-hour (DLH).

In this example, there are two manufacturing departments and four service departments. Likewise, the procedure for allocating costs in stage 1 can be considered correct. Nonetheless, if improper allocation bases were used in stage 2, the product costs would still be distorted.

Using DLHs as the allocation basis, we would calculate the application rate by dividing total departmental overhead by the number of DLHs. When calculating the cost standard for products P_1 and P_2 using the application rate of $280 per DLH, the allocated amounts for P_1 and P_2 are calculated using the application rate times the respective DLHs 1 and 9. The overhead cost allocated to each unit is then calculated by dividing the allocated amounts per DLH by the production rate per DLH (10 units and 100 units respectively). As a result, the overhead allocated is $28.00 per unit of P_1 and $25.20 per unit of P_2. This process is depicted in Table 5-1(A).

Table 5-1 (A). Traditional Method (Direct Labor-Hours)

Overhead rate = Overhead ÷ direct labor-hours
= $ 2,800 ÷ 10
= $ 280/hour

Amount of allocation
P_1 = $ 280/hour × 1 = $ 280
P_2 = $ 280/hour × 9 = $2,520

Overhead cost per unit
P_1 = $ 280 ÷ 10 = $ **28.0/unit**
P_2 = $2,520 ÷ 100 = $ **25.2/unit**

The Multiple Basis Volume-Driven Allocation Method

In the 1980s, many Japanese companies responded to the new manufacturing environment by using machine-hours (MHs) in addition to direct labor-hours in stage 2 of their allocation system. Machine-hours are naturally a more appropriate measure than direct labor-hours in a machine-paced manufacturing setting.

Some Japanese companies use more than two volume-related allocation methods simultaneously. However, an approach that uses two volume-related allocation methods—a DLH overhead rate and a MH rate—is illustrated. With this approach, machine processing costs are allocated to a separate cost center and charged to products on the basis of MHs. Other overheads are allocated by using DLHs. The two overhead rates in our example are $225 per DLH and $50 per MH.

As the computation process shows, the unit costs for overhead are calculated as $27.50 for P_1 and $25.25 for P_2. Allocation amounts would be calculated as shown in Table 5-1(B). Readers will notice that the computation result is not substantially different from the first method results.

Table 5-1(B). Multiple Volume-Based Allocation Method

Direct labor-hours

Overhead rate = Overhead ÷ direct labor-hours
= \$ 2,250 ÷ 10
= \$ 225/hour

Amount of allocation
P_1 = \$ 225/hour × 1 = \$ 225
P_2 = \$ 225/hour × 9 = \$2,025

Machine hours

Overhead rate = Overhead ÷ machine hours
= \$ 550 ÷ 10
= \$ 50/hour

Amount of allocation
P_1 = \$ 50/hour × 1 = \$ 50
P_2 = \$ 50/hour × 10 = \$ 500

Overhead cost per unit
P_1 = (\$ 225 + \$ 50) ÷ 10 = \$ **27.50/unit**
P_2 = (\$2,025 + \$500) ÷ 100 = \$ **25.25/unit**

Activity-Based Costing (ABC)

ABC is based on the assumption that what traditional systems call overhead costs incur because some resources perform activities not related to volume. Allocation bases in our hypothetical example are:

1. (for design costs) the number of design changes
2. (for setup costs) the number of setups needed
3. (for purchasing costs) the number of orders
4. (for material-handling costs) the number of parts

While these four examples clearly oversimplify activities that give rise to specific departmental costs, the purpose of this presentation is to give an example of ABC rather than to consider the most appropriate cost driver. In practice, activities are categorized into the unit, batch, product, and facility-sustaining taxonomy for activity-cost drivers. Table 5-1(C) shows the computation of overhead using an ABC approach.

Table 5-1 (C). Activity-Based Costing

Computation of Overhead Rate

Activities	Cost Drivers	Overhead		Assignment Bases	Overhead Rate
Assembly	Direct labor-hours	$ 800	÷	10 DLH	= $ 80/DLH
Machine work	Machine hours	$ 550	÷	11 MH	= $ 50/MH
Design	No. of designs	$ 420	÷	2 designs	= $ 210/D
Setup	No. of setups	$ 270	÷	3 setups	= $ 90/S
Purchasing	No. of orders	$ 240	÷	4 orders	= $ 60/0
Material handling	No. of parts	$ 520	÷	10 parts	= $ 52/P

Computation of Allocation Amount

Activities	P_1	P_2	Total
Assembly	$ 80 × 1 = $ 80	$ 80 × 9 = $ 720	$ 800
Machine work	$ 50 × 1 = $ 50	$ 50 × 10 = $ 500	$ 550
Design	$ 210 × 1 = $ 210	$ 210 × 1 = $ 210	$ 420
Setup	$ 90 × 1 = $ 90	$ 90 × 2 = $ 180	$ 270
Purchasing	$ 60 × 1 = $ 60	$ 60 × 3 = $ 180	$ 240
Material handling	$ 52 × 3 = $ 156	$ 52 × 7 = $ 364	$ 520
Total	$ 646	$ 2,154	$ 2,800
Volume of production (units)	10	100	110
Overhead cost per unit	$ 64.60	$ 21.54	$ 25.45

This example shows the assigned overhead for products P_1 and P_2 calculated using ABC. Table 5-1 makes it clear that the cost of P_1 (a product produced in small lots of numerous types) is in fact significantly higher than the cost of P_2. Conversely, the traditional method of volume-related costing makes the cost of P_2 (mass-produced products) appear higher.

In this example of product costing it becomes clear that in the stage 2 procedure for assigning costs, traditional volume-related bases over-allocate costs to mass-produced units. This occurs because more non-volume-related activities are required when producing a variety of different products in small lots than when producing large quantities of only a few products. Therefore, when the application basis is volume-related, the mass-produced products (P_2) appear to subsidize those produced in small lots (P_1) by absorbing much of the non-volume-related cost.

As we can see, ABC is a theoretically sophisticated tool for assigning factory overhead to a cost objective, most commonly a product or product line. It accomplishes this partly by redefining the problem into one of "driving resource costs to activities performed." Under the traditional method of assigning overhead, mass-produced products were made to bear higher costs because the difference in relative amounts of non-volume-related overhead consumed were not taken into account. ABC acknowledges the difference in the relative amount of investment for each cost objective and assigns an appropriate amount to each. ABC also acknowledges various sources of costs or cost drivers (such as setup, material handling, design, and booking of orders). Overhead is then assigned to each product using measurements or estimates of cost-driver rates. With ABC, companies can engage in sophisticated product profitability analysis. In addition, cost analysis using ABC can help companies understand cost reduction possibilities by providing a data base for estimating the savings from certain activities.

How to Install ABC

To install a transaction-based ABC system, two alternatives exist: (1) to revamp the current cost accounting system or (2) to build an entirely new system.

Kaplan's Arguments

In the early stages of ABC development, Kaplan (1988) argued for separate cost accounting systems for different purposes. His reasoning was based on the three different functions assigned to cost accounting: inventory valuation, operational control, and product cost measurement.

No cost accounting system can cover all three functions by itself and managers are dissatisfied with those that only prepare financial reports and value inventory. Kaplan rightly argues that computer development has made it economically feasible to gather, process, and report information using multiple cost accounting systems at the same time. He further advocates that a company should have multiple systems.

Kaplan (1990b) states that many financial executives have little enthusiasm for his recommendation. Recalling the four stages in the evolution to ABC, he suggests that in the future companies with multiple cost systems will need to move to integrated costing systems (stage 4).

Stage 1's cost accounting system is found in small companies. The stage 2 cost accounting systems are used primarily to prepare public financial reports. This is the current stage in many U.S. companies. Stage 3 is the multiple cost system approach that Kaplan advocated earlier. Lastly, stage 4 utilizes two integrated systems: one ABC system for evaluating the profitability of products and clients, and another for on-line feedback and performance evaluation. Periodic financial report information is extracted from both systems and reports are prepared to fulfill externally determined standards. As a further development of this idea, numerous companies are beginning to experiment with using the newly devised ABC system for operational control. However, compromises are made in an attempt to perform both managerial functions—operational control and strategic profitability—with a single system (Kaplan 1990b).

Arguments for Multiple Cost Systems in Japan

What was the Japanese response to multiple cost systems? The same line of argument as Kaplan's was raised by academics in the 1970s (Itoh 1978; Kobayashi 1978; Sonoda 1979). They argued that rapid computer development made it possible, perhaps even mandatory, for companies to have multiple cost accounting systems targeted at different activities (for example, financial statement preparation, cost control, profit planning, and decision making). The goal was to add flexibility to the cost accounting system. However, these concepts were not favorably received by industrial accountants of the day.

The author sought views of Japanese accountants toward Kaplan's arguments at the April 1988 meeting of the Managerial Accounting Research Group of the Tokyo chapter of the National Accounting Association (IMA today). Negative views predominated at that time: "Having three cost accounting systems may cause us to lose sight of which cost to consider as the true product cost," "Despite how computers have developed, multiple cost systems will still be very expensive," or "We are satisfied with our present cost accounting system."

Then, in October 1988 the author conducted a mail survey on this issue. It revealed that, while many still opposed having multiple cost accounting systems, as many as 27 percent of Japanese companies already had them.

What about current Japanese cost accounting systems? While they appear to be simple, they actually are more advanced than those depicted in Kaplan's stage 2. This is due to the influence of 1962's Cost Accounting Standard that aimed to achieve cost control and budgeting goals as well as aid the preparation of financial statements (Cost Accounting Standards, 1.2., 1962). The fact that only 27 percent of Japanese companies report having multiple costing systems probably understates the usefulness of Japanese planning and control systems since Japanese single costing systems are more oriented toward planning and control functions than Kaplan's stage 2.

The cultural bias in the four-stage framework, although subtle, causes significant problems in understanding Kaplan's concept of cost measurement. This may even be true in the United States, although to a lesser degree, where we expect some companies to have a stronger bias toward planning and control than that suggested by the four-stage model. This difference in the basic philosophy driving the accounting systems may be one factor leading to the cool response of Japanese managers to ABC as opposed to the enthusiasism of American accountants.

ABC versus Target Costing

To clarify the characteristics of original ABC, we will compare it to target costing. The two techniques, both typical in their respective countries, can be differentiated as follows:

1. ABC focuses more on support costs.
2. ABC is more a financial tool than an operations tool.
3. ABC's main contribution has been to product profitability analysis rather than cost reduction.

ABC's Focus on Support Costs

ABC uses direct cost information in the same way as the traditional system. Where ABC differs is that it creates a bill of activities for overhead (Sharmen 1990). More concretely, ABC focuses on such support costs as manufacturing overhead, marketing costs, and other corporate overhead in an attempt to assign them to products by tying the assignment more tightly to the activities that lead to spending. Thus, ABC focuses on the skyrocketing support costs so common as a result of product diversification and factory automation.

Target costing likewise deals with support costs. However, in most companies its main focus historically has been to reduce direct materials cost. In the Japanese cost management approach, a different part of the system, kaizen costing, focuses on reducing such support costs as setup, design, purchasing, quality control, receiving materials and parts, and other engineering support costs as well as direct material costs. In kaizen costing, these support costs are controlled by the nonfinancial system and budgeting in conjunction with engineering tools like VE, TQC, or JIT.

ABC Is a Financial Tool

ABC provides cost information that can be used for product costing, cost control, and other management purposes. This makes it more than a general ledger system. However, in its original concept, ABC is an accounting or financial tool rather than an engineering one. For example, ABC is "a new accounting system for better product cost" (Campi 1992). And many accountants consider it to be a cost assignment tool (Turney 1992). In fact, when Kaplan first encountered ABC systems in the mid-1980s he described them as an allocation procedure by which operating expenses were assigned, via activities, to products and services (Kaplan 1992). (Later works describe the process as an estimation process and not an allocation process. These writings are part of the movement toward ABM and are discussed below.) In fact, ABC is a better technique for overhead assignment than traditional volume-related allocation methods and can improve a company's performance measurement systems. As such, its primary output is financial information for use in decision making.

On the other hand, target costing is a goal-oriented decision technique, or process. Its major focus is to manage the design of products for manufacture. Target costing provides the decision environment in which any relevant information (including ABC information) can be used. Instead of accounting techniques, engineering methods are emphasized. The use of VE is inevitable with target costing and TQC, JIT, TPM, or other engineering techniques are often installed concurrently.

Product Profitability Analysis

The main goal of ABC is to provide information for a variety of purposes, most commonly product profitability analysis for pricing or product mix decisions. Cooper and Kaplan (1991) view improved decisions as an important benefit of ABC by stating that "the more accurate product costs reported by activity-based cost systems reduce the possibility of managers making poor decisions based on available cost information." In fact, ABC literature sometimes creates the myth of a "super secret X-ray viewer" that compels managers to make better decisions. ABC is also useful as a tool for continuous improvement in that identifying and costing activities can provide powerful information for cost reduction. However, this use is more difficult to show (and has a lower *gee-whiz!* value).

For these reasons, or perhaps because of a genuinely different focus, ABC's cost reduction function received significantly less coverage than the strategic price and mix decisions. Johnson (1992) has commented that activity-based cost management tools "did not generate process maps, had no customer focus, and did not lead to bottom-up ideas for generating continuous process improvement." In fact, ABC provides information for process improvement—but not for improving processes by ABC itself. The ABC project "must have a top-down commitment to succeed" (Norkiewicz 1994). Since ABC sometimes lacks customer focus, some Japanese companies use value analysis (VA) with ABC to maintain customer satisfaction.

In contrast to ABC, target costing has little to do with the product costing system. Target costing is a process for strategic cost management. Actually, most Japanese managers do not place much importance on the accurate overhead assignment advanced by ABC proponents. What they want to know is how to reduce overhead by reengineering production processes. In short, target costing as well as kaizen costing are typical tools for producing quality products with low costs—including overhead. It emphasizes a process philosophy rather than a measurement philosophy. For target costing to be useful for cost reduction, target costing must have a "bottom-up commitment to succeed." ABC provides inputs to a decision technique for improving the use of current and anticipated resources; target costing is a tool for changing the nature and magnitude of currently available resources.

This distinction may be part of the development of our respective countries, particularly after World War II. The overwheming majority of business methods developed or implemented in the United States in the postwar period were designed to optimize the deployment of resources in an equilibrium condition (i.e., most IE/OR methods including linear programming). Even when such an optimal selection assumption is not in the method itself, it may be in the users' minds. The EOQ model story is a case in point.

EOQ itself can easily be used in the JIT environment even though it tended to hide the fact that ordering and holding costs are long-term variable costs and not externally fixed. For reasons we cannot explain, the Japanese seem to have freed themselves of many of these ties; they see not the equilibrium, but the flux. Many of the three-letter Japanese methods can be viewed as based in such a flux management paradigm.

ABC's Impact on Japanese Accountants

Interestingly, ABC did not gain much popularity among Japanese accountants until mid-1991, when the recession in Japan began. However, many Japanese companies have felt compelled to restructure their organizations and reengineer their business processes since late 1991. When Robin Cooper and this author visited Japanese companies to conduct research on overhead management in August 1989, no Japanese manager showed any interest in ABC. They wanted to successfully manage overhead—not compute an accurate product cost. This was further confirmed in a January 1991 mail survey. In the survey, over half of the respondents had not heard of ABC. Of those who knew of it, a third simply ignored it. Three more were considering its use. Only six companies (4 percent) used ABC; they all modified it to their needs and had strong ties to U.S. firms.

Empirical surveys show that British firms exhibit great reluctance to change from their traditional cost accounting systems to techniques such as ABC (Bromwich and Bhimani 1989). However, Bailey (1991) reported several implementation examples of companies using ABC. Morrow and Ashworth (1994) mentioned that ABM applications have grown well beyond those used for product costing through the increasingly varied use of activity analysis techniques.

Why were Japanese managers less interested in ABC than American accountants? Of the many possible reasons, four stand out. (1) A Japanese innovation in costing called Direct Charge of Overhead to Product Line System (DCOPLS) gained momentum (see Chapter 4). (2) The Japanese focus on cost reduction more than on cost deployment. (3) The Japanese appear to want the flexibility to over- or under-cost for tactical management purposes (e.g., to manage employee behavior). (4) The original presentations of ABC were firmly embedded in the particular social and economic situation existing in the U.S. at that time.

Direct Charge of Overhead to a Product or Product Lines (DCOPLS)

The costs of support departments are commonly allocated in the United States. However, Japanese companies typically assign or trace them to a product or product lines where possible; that is, they already have practices very similar to ABC embedded in their traditional systems. Thus, the improvement in measurement from using ABC, although it exists, is not as compelling for Japanese companies as it is for U.S. companies.

For example, setup costs are usually charged to products and are treated as direct costs. Design costs are also typically assigned to a product. At Epson, setup costs, material-handling costs, inspection costs (except quality assurance), and maintenance costs are treated as direct costs. When design costs are treated as overhead, some companies use a weighted application to cope with this distortion. Fuji Electric, for example, uses a weighted rate using activity-related bases to correct the distortion caused by using volume-related allocation bases. The product costing practices of another major company that produces and markets telecommunications equipment, information-handling equipment, electronic equipment, and electronic parts are shown in Table 5-2.

Table 5-2. Typical Japanese Treatment of Overhead

Cost Element	Low Volume	High Volume
Design cost	Direct charge	Allocation
Material handling	Direct charge	Direct charge
Setup	Direct charge	Direct charge
R&D costs	Allocation	Allocation
Quality control	Direct charge	Direct charge

This type of strong push to cost tracing reduces the size of the costs allocated, thereby reducing the distortion caused by the allocation process. All of this occurs within the bounds of what appears to be a traditional volume-based system and indicates that some of the failures seen in Kaplan's two-stage systems could be design and implementation failures rather than method failures.

Cost Reduction versus Product Mix Decisions

Although ABC's genesis may have been the desire of design engineers for better information to use for different management purposes, the goal of its original form was to improve profitability by having more accurate product costs. These accurate costs can be used effectively to restructure an unprofitable product mix. Product mix, along with pricing, is one of the strategic areas with the most observable short-term results.

Japanese managers believe that competition between companies in Japan is more intense than in the United States. This leads them to intense cost-reduction efforts. In contrast, because American stockholders wield such power, their management has focused a lot of effort on increasing measured profitability by deploying and redeploying resources. Moreover, U.S. companies in the late 1980s were struggling to restructure their business organizations. This is quite similar to the current situation in Japan where companies are suffering from a depression and need to restructure their business organizations. ABC is an effective way to identify unprofitable products. In its original form, ABC is the most powerful and sophisticated tool known for analyzing product profitability for redeployment purposes (with the strongest short-term impact on the financials). However, it is less effective for reducing costs in this context since reductions in consumption have a looser link to financials.

Fostering Product Variety with Volume Production

American executives who believe that producing a variety of products in small lots is laborious and costly might think it natural to assign to these lines a greater burden of costs by using ABC. They then might try to eliminate the low-volume unprofitable operations and increase profitability, ROI, and stockholder dividends.

In contrast, the Japanese sense of values is becoming increasingly diverse, and types of products are becoming more varied, so that the best way to cope with consumer satisfaction is to produce a variety of products in small quantities. Thus, it may be an appropriate accounting strategy to subsidize high variety, low volume production by assigning some of the high cost to mass-produced products. Because the lot size of a newly introduced product is always small, to burden it with high costs reduces the incentives of mid- and low-level managers to develop new products. Of course, if allowed to continue unchecked, this can lead to excessive variety and model changes. Having witnessed this phenomonon in the strong business climate of the late 1980s, Japanese automakers recently attempted (some without success) mass-customization as their bubble economy collapsed (Pine, Victor, and Boynton 1993).

Japanese managers typically value long-term relationships with clients. As a result, were a product to become temporarily unprofitable or significantly less profitable, they would be unlikely to drop their customer. Hence, ABC would compromise their ability to tactically over- or underprice products and provide information for types of decisions they would not normally contemplate during healthy economic periods.

Social and Economic Factors

In Japan, managing for long-run competitiveness has emphasized continuous improvement. Thus, Japanese managers have shown little interest in revising the cost accounting mechanism to focus on the product. Instead, by directly assigning factory overhead, they have turned their attention to analyzing raw data with the intent to reduce costs. In fact, while a 1988 survey showed 50 percent of U.S. managers listing "developing a selected basis for overhead allocation" as their top priority, the first and second choices of Japanese managers were "moving to variable costing" (36 percent) and "minimizing factory overhead allocation." This was out of six items to be selected (NAA Tokyo Chapter 1988).

As a tool for restructuring business organizations, ABC is very effective. However, until mid-1991, Japanese managers were totally disinterested. After all, who thinks about using cost accounting tools to restructure one's product mix when business is good! Let's look at some typical reactions of Japanese managers toward ABC from the late 1980s to early 1991:

1. The controller of Nippon Steel believed that his company was assigning overhead with a method similar to ABC. Their foundries could not be successfully controlled using only a volume-related allocation bases. He added that the only difference between ABC and Nippon Steel's system was that ABC was overly complex.
2. Komatsu's R&D director expressed concern that producing a variety of products in small volumes would involve a larger assignment of costs than at present if Komatsu were to use ABC. It might therefore interfere with new product development.
3. A cost accountant at Clarion argued against introducing ABC by saying that they could not cope with the diverse demands of their customers and produce in small quantities. He added that the original ABC opposed their goal of customer satisfaction.

These comments may be difficult for U.S. observers to understand. What, after all, is the mission of the organization if not to make a profit? As we examined in Chapter 1, the prime mission of Japanese management revolved around organizational survival. The Japanese interpret the word "organization" very differently than do most Americans. They usually consider the organization as (in order of importance) employees, suppliers, customers, banks, management, society, and maybe stockholders. The company itself as a legal entity does not figure strongly in this hierarchy.

To many U.S. managers, organizational survival may relate more strongly to the legal entity than to the employees. Of course, this is how they justify large-scale layoffs—as a way to preserve the organization. The Japanese perspective views layoffs not as preservation but as the destruction of the organization. Traditionally in Japan, profit is less important than the physical integrity of the company. This is true even at the highest levels of management. In such a situation, to discuss methods (like ABC) whose main selling point is the possibility of damaging these relationships is heretical unless the threat of catastrophe is severe. In addition, Japanese managers seem willing to decrease short-term profitability in exchange for stable relationships that they view as more instrumental for survival. This traditional Japanese perspective is changing now partly due to the severity of the economic recession and partly due to the advent of a new generation. Nonetheless, values so deeply ingrained make this change extremely problematic.

Activity-Based Management (ABM)

Discussions surrounding ABC have shifted from a concern with product costing to concern with process improvement. For example, in the early stages of selling ABC, Cooper wrote about it as a more accurate fully-allocated unit cost system (Cooper 1988-1989). A recent paper, however, is devoted to activity-based management or ABM (Cooper, et al. 1992). Kaplan (1992) stated that when he and Robin Cooper first encountered ABC systems in the mid-1980s at sites such as Schrader-Bellows, John Deere, and Union Pacific Railroad, they described what they saw as an allocation. Their later writings move away from viewing cost assignment as allocation and toward viewing it as estimation. In short, many ABC researchers have shifted their focus from measurement to process, and from product cost analysis to process cost reduction. This has sometimes changed the name to ABM or activity-based cost management (ABCM).

What Is ABM?

ABM refers to the use of ABC to help an organization improve the value of its products and services (Raffish and Turney 1991). As its major information source, it draws on activity-based analysis (ABA)—not on ABC per se (although obviously ABA often leads to ABC). This discipline includes cost-driver analysis, activity analysis, and cost reduction.

Cost Reduction Through ABM

ABM has two major aims (Turney 1992). One is to improve the value received by customers. The other is to improve profit by providing this value.

The first step in implementing ABM is to analyze activities. This accomplishes the following:

- identifies nonessential or non-value-added activities
- analyzes significant value-added activities
- compares activities to the best practices
- examines links between activities

The second and crucial step of ABM is to practice cost-driver analysis to reduce cost. The best way to reduce cost is to change the way activities are used or performed and then to redeploy the resources freed by the improvement. Five guidelines suggested by Turney (1992) are:

1. Reduce time and effort.
2. Eliminate unnecessary activities.
3. Select low-cost activities.
4. Share activities wherever possible.
5. Redeploy unused resources.

According to Monden (1993), the practices of ABM are exactly the same as kaizen costing in their aim—continuous operational improvement—practiced by major Japanese companies.

The Need for Paradigm Shift

This shift from ABC to ABM heralds a favorable change to business management for American and Japanese managers alike. Because of this change, many Japanese managers have recently become interested in ABC or ABM. However, to introduce ABM but not ABC, the cost management view must dominate the business organization. Although calling this a paradigm shift may be an exaggeration, it would be an extremely pronounced departure from prior practices.

Keating and Jablonsky (1990) argue that in the predominant paradigm, which prevails in typical American firms, most financial communication flows vertically up through the chain of command. The systems that produce financial information also tend to be divorced from systems that produce operating information. In this measurement regime, senior management's role is to make the major resource allocation decisions, with the help of staff and external experts whenever necessary. Management's major task is thought to be to provide an orderly assimilation, exploitation, and coordination of separate sources of purchased expertise.

Performance evaluation through financial systems is necessary in this type of system, which has been called the command-and-control philosophy. The smoke stack metaphor is often used by critics of this approach because the controlled upward flow of smoke in separate chimneys is an appealing visual example of both the separateness of the information from other information as well as the bottom-up flow. Both features—separation and vertical movement to the outside—seem to be inherent in command-and-control organizations. Because of the arm's length professional orientation to the financial communication act, measurement plays an important role in this type of organization. The original ABC is probably the most sophisticated and effective accounting tool for this type of management.

A comparative study of strategic management practices in large U.S. and Japanese firms (Kagono et al. 1985) concluded that the strategic orientation of U.S. firms emphasizes the mobile deployment of financial resources. In Japanese firms, the strategic orientation tends to be expressed in terms of long-term accumulation and deployment of human resources. In the same study, Japanese executives reported a significantly less elaborate set of formal financial control systems than reported by American executives. Japanese managers rely more on employees' ability to learn about the productive process and to make decisions about process improvements based on their "first-order observations." These views are confirmed by an examination of Japanese cost-accounting systems, where there was much less evidence of the highly technical and financial accounting tools of a command-and-control system (Scarbrough, Nanni, and Sakurai 1991).

When we discuss a "process approach" we mean managing a business based on a philosophy of reducing costs through process improvement. To reduce cost through process improvement primarily involves people in the production and engineering field, albeit with assistance from support areas such as accounting. This management mode of operation is coming to be known as a competitive-team philosophy. Financially-oriented managers (particularly former CPAs in controllership positions), which predominate in the United States, would need what amounts to a paradigm shift in management philosophy to install ABM in a manner consistent with its basic assumptions. We suspect that many ABM systems developed in America will fail to achieve their targets if they install only its form and not ABM's underlying philosophy. Tell-tale signs of the deployment perspective are easy to spot. For example, an orientation toward a formal monthly closing of the books can persist only in a measurement-oriented organization.

ABM Case Studies in Japan

Japanese management began to show an interest in ABC/ABM in late 1991 when their companies started to suffer from a deep recession. Since then, major Japanese companies have experienced a round of restructuring in an attempt to strengthen their business operations. However, upon realizing that traditional restructuring was not enough to cope with the competitive markets, managers began to reengineer their business processes. Like their American counterparts, they chose ABM as one tool for reengineering the business process (Yoshikawa 1994; Sakurai 1994; Hiraoka 1994; Takayanagi 1994). In fact, major Japanese companies began reengineering using ABM. The following two examples are typical of ABM in Japanese companies.

Case Study: ABM in Omron

Omron's corporate strategy is to transfer low-tech products to overseas operations and to produce high-tech products at home. Typically, low-tech products are mass-produced overseas because of lower labor costs. The wide variety of high-tech products are generally produced in small lots. Sales figures for these high value-added products are dropping as the marketplace grows more competitive.

The three plants in the company's Mishima division make different products: FA systems, information devices, and test systems. The information-devices plant manufactures computer input devices and peripheral equipment. This plant is highly automated with only 320 employees (as of January 1994). However, the marketing area is growing uncompetitive due to the high value of the yen, Japan's poor economy, and the resulting decrease in demand domestically for office-automation equipment. In addition to the loss of overseas customers due to the high yen, Omron's major clients have established plants overseas and are buying from foreign suppliers. Consequently, sales volumes and profits are falling yearly. In response, company management has decided to reengineer its business processes with the goal of making a profit even at the lower sales volume.

Omron's use of ABM

To undertake business reengineering in the information-devices plant, all managers wanted to know their actual product costs.

This led to the use of ABM. When Omron implemented ABM in 1993, the information-devices plant had 367 employees. The ratio of direct labor to support staff was 25 to 75 (or one-third). The rate of outsourcing was quite high and production was wide variety/small lot. This made support costs such as material handling and shipping of ordered products very high in comparison to other companies.

It took three months (May to July 1993) to prepare the activity-based analysis needed for ABM. Two full-time and six part-time employees were selected from different departments to undertake the analysis. In addition, when necessary, 30 hourly employees helped the ABM team analyze activities and gather cost data.

The core of the analysis project was to identify and classify business activities. Of 180 activities identified, 40 were non-value-added; for example, information forecasting (twice a month), plan modification, and work flow management (e.g., inventory management). The output of the project was a comprehensive description of the organization's cost flows linking activity costs to products and customers. This was like a snapshot of the firm, and no attempt was made to institutionalize the system for gathering this information. However, the learning and management insights were institutionalized by changes in the company's training materials and procedures. This allowed the process of managing and discussing alternatives with an activity-based perspective to become the norm, even though actual record keeping has not followed suit.

Results of applying ABM

As a result of its activity analysis, Omron discovered that costs driven by product changes were a primary factor depressing profitability. Management was surprised to learn that these costs were a significant portion of total costs and decided to focus its cost reduction program on these activities. The initial analysis revealed two main types of change to study: engineering changes (technology driven) and product changes (customer-demand driven).

Engineering changes are driven by innovations in the production process or underlying technology. Product-planning changes are in response to customer needs or market preferences.

Direct costs can be easily charged to a product. However, indirect costs are assigned to each product line. First, plant-produced products were classified into eight categories and then further classified into sixteen subcategories. (Because some products were not produced, there were only 80 subcategories.) Thus, the company classified indirect costs into 80 product lines without allocating the costs to each product. Analysis disclosed that only 20 percent of the product lines produced were profitable. The greatest losses occurred in the prototype products; they cost four times the selling price.

Advantages gained from applying ABM were the following. First, ABM disclosed unprofitable product lines. Second, it made measuring "true" costs possible. (The plant's three types of products in descending order of profitability were: mass-produced products (low variety/high volume), small-lot products (high variety/low volume), and unique prototype products.) Third, as it turned out, technological changes and product-planning changes were the major causes of low profitability. As a result, management instituted a cost reduction program on the shop floor, leading to a reduction of 47 jobs. These 47 workers were rotated to other Omron plants. The result was an improved ratio of shop floor workers to support workers of 40/60 (or two-thirds).

The ABM program was promoted and executed by the production management department, not by the financial accounting or cost accounting department. Cost data by activity was prepared by production management. However, one team member belonged to a cost management group and had a cost accounting background. The ABM project team also used cost data provided by the cost accounting department. This program was integrated with the accounting data so that the financial bottom line was the same as the cost accounting data.

The next round of ABM analysis was in late 1994. At that time, ABM was extended beyond manufacturing to the marketing area as was expected.

Case Study: ABM in Sanyo

Sanyo Electric introduced ABM as the main driver for reengineering its business processes in 1994; ABM is used as a tool for process improvement and innovation.

Sanyo Electric is a major consumer electronics manufacturer; it has about 15,000 employees. Sanyo's air conditioner division (which has about 1,000 employees) began using ABM in 1994. Exports have been as much as 50 percent of production in the past, but have declined to 30 percent; in other words, its markets are now mainly domestic. Other Asian countries provide the strongest competition in the consumer product industries; Japanese companies—including Sanyo—are struggling to survive in these industries. Even in the domestic market, competition is keen among Japanese manufacturers.

The overhead-to-manufacturing-cost ratio has increased greatly at Sanyo. Using the ratio for 1989 as the basis (100 percent), by 1993 the ratio had risen to 151 percent. The most striking increase was in indirect labor, which rose 157 percent during the period. This rapid increase can be linked to the active expansion program established in the mid-range business plans during the bubble economy period. For example, Sanyo's expansion of plants, equipment, and workforce caused much of the increase in overhead. As part of its fight for survival, Sanyo introduced ABM to reduce overhead to the appropriate level.

Aim of introducing ABM

Restructuring and reengineering have a dark side—layoffs, plant closures, and product pruning. Because of this, many employees are reluctant to participate in ABC/ABM programs. In some cases, participants who promoted ABC were the ones fired. In fact, employees often do not promote ABC out of fear about their jobs. (For example, employees in a Japanese-affiliated company in the United States were reluctant participants in ABC because the employees saw several workers lose their jobs or status.)

Thus, the goals for ABM at Sanyo included promoting new businesses, not just restructuring or reengineering; this broader aim can motivate employees to participate. In fact, Sanyo employees proved willing. The net result of the program was a reduction in labor needs of 137 field workers and 115 support workers, out of 1,000 employees. Most of the spare workers will be rotated to new businesses in the future.

To support the development of new businesses, the R&D department was restructured. Sanyo took three measures to insure this step:

1. A New Business Development Committee meeting was held once a month to find and discuss promising new businesses.
2. The R&D department, which has previously had a mixed mission of doing both basic and applied research, was turned into a New Product Development Institute whose goal was solely to develop new products.
3. The decision was made to move spare employees into new businesses. But, since it is impossible to rotate spare field workers to new businesses unless they are prepared, Sanyo established an in-house training center to educate these employees for up to three years.

Design of the ABM systm

Sanyo has a standard cost accounting system. However, it expects the system to have little or no cost-control function, which may contrast with how standard costing is used in some Japanese process-oriented industries. Sanyo's standard cost accounting system remains extremely useful for preparing financial statements; it is integrated with a host computer and functions well.

Sanyo decided to have its ABC system operate independently from the standard cost system on the host computer. It was thought that the ABC system on a personal computer would be effective.

Sanyo inaugurated its ABM program in January 1994. The system design took from January to March, and installation from May to July, 1994. There were twelve employees involved in the

ABM program, but none exclusively. The results were presented at a special meeting of top and middle management.

There were 150 cost drivers identified in all. For example, there were 16 cost drivers in the planning and design process, 7 cost drivers in the production designing process, 38 cost drivers in the production control process, and so on. This set of cost drivers was sufficient for providing information to top management, though more detail would be needed for process improvement at lower levels. For that reason, the company intends to expand the number of cost drivers and analysis to include general and administrative expenses. Sanyo will have ABC budgeting in 1995.

Benefits gained from ABM

The biggest benefit gained from introducing ABM was that Sanyo found many cases of non-value-added expenditures, most of them attributable to the "fat" gained during the bubble period. This fat included many underutilized resources in the company, but without an activity analysis perspective, the underutilization was not apparent. ABM helped Sanyo detect that more than one-fourth of the employees in the division were not adding value. This knowledge contributed to effective management.

The concrete benefit gained from introducing ABM was that Sanyo identified 252 employees out of 1,000 as unnecessary. One of the striking characteristics in Sanyo's ABM system is that the company rotates those spare employees to new value-added businesses, with no layoffs. This no-layoff policy helped motivate employees and made the program a great success.

Sanyo also identified three problems:

1. Barriers between departments
2. Backward management of human resources
3. Lack of firm business plan for new business development

The president of Sanyo asked divisional managers to work on reengineering solutions to these three problems in the reengineering program that began in late 1994.

Conclusion

ABC analysis supplies information useful to ABM which, in turn, can be a powerful tool for continuous improvement and process reengineering. ABC emphasizes cost measurement while ABM focuses on process improvement. ABC can provide valuable information on product profitability analysis, customer relationships, product design, process improvement, and supplier relationships by assigning factory overhead to activities and then to products. If used to aid in ABM, ABC is akin to what most Japanese companies practice as kaizen costing; it can be extremely effective in reducing overhead. Since target costing can successfully reduce direct costs (such as material and parts), these three can complement each other as Table 5-3 shows.

Table 5-3. Relationship of ABC, ABM, and Target Costing

Tools	Main Purpose	Cost Elements	Emphasis
ABC	Product profitability analysis	Overhead	Cost assignment for managerial decision making
ABM	Process reengineering	Overhead and direct costs	Process improvement
Target Costing	Strategic cost management	Direct costs and overhead	Cost reduction

Because of its similarity to kaizen costing, we do not expect ABM to become as popular in Japan as it is in the United States. However, we think it will influence the development of kaizen costing and, in the process, perhaps produce some unique additions to the body of cost management knowledge.

Japanese managers have shown a greater interest in ABM as a tool for business process reengineering since late 1991. The severe recession in Japan has forced many companies to engage in a fierce reengineering effort, and many see ABM as a tool to aid this process. ABM may also help to reduce the labor-hours in support work. This is crucial in Japan where a negative population growth continues to reduce the number of workers entering the economy. In addition, some automobile and home electronics companies are examining the original formulation of ABC in hopes of reducing the variety of models and number of parts and thereby reducing costs.

Since the activity analysis that underlies ABM is similar to that underlying kaizen costing, the latter's analytical data base can be integrated with ABM with relatively little adjustment. This allows the running of a relatively lower-cost kaizen costing system along with the use of activity analysis for ABM without drastically changing the costing system or increasing accounting costs.

Our general conclusion is that there is parallel progress in the United States and Japan toward the goal of creative management of overhead consumption. Looking under the surface, we see that the bulk of the effort is very similar even though the language and presentation differ. The American contribution is more precisely articulated and more theoretically advanced, although it embodies several culturally specific assumptions about employee knowledge and training and organizational goals that are inappropriate in Japan. The core similarity is that both sets of practice attempt to assign costs based on resource consumption. The ABC concepts are having a strong effect on Japanese management practice by leading to a refinement of the existing approaches—kaizen costing in particular. We should emphasize that the main effect is on analysis and the understanding of costs; there is a much lower impact on the design of the transaction-capturing systems, which they intend to keep simple.

References and Further Readings

Bailey, Jim. 1991. "Implementation of ABC Systems by U.K. Companies." *Management Accounting* (February) p. 30.

Berliner, Callie, and James A. Brimson. 1988. *Cost Management for Today's Advanced Manufacturing.* Boston: Harvard Business School Press.

Brimson, James A. 1991. *Activity Accounting.* New York: John C. Wiley & Sons pp. 69-70.

Bromwich, M., and A. Bhimani. 1989. *Management Accounting: Evolution Not Revolution.* The Chartered Institute of Management Accountants.

Campi, John P. 1992. "It's Not as Easy as ABC." *Journal of Cost Management* (summer) p. 8.

Church, A. Hamilton. "The Proper Distribution of Establishment Charges." *The Engineering Magazine* Vol. 22, no. 2, pp. 231-240.

Cooper, Robin, and Robert S. Kaplan. 1988. "How Cost Accounting Distorts Product Cost." *Management Accounting* (April) p. 22.

Cooper, Robin. 1988-89. "The Rise of Activity-Based Costing: Parts 1, 2, and 3." *Journal of Cost Management* Vol. 2, no. 2; no. 3; no. 4. These were written from September 1988 to Winter 1989.

Cooper, Robin. 1989. "The Rise of Activity-Based Costing (Part 4): What Do Activity-Based Cost Systems Look Like?" *Journal of Cost Management* (Spring) p. 38.

Cooper, Robin. 1990. "Cost Classifications in Unit-Based and Activity-Based Manufacturing Cost Systems." *Journal of Cost Management* (fall) pp. 4-14.

Cooper, Robin, and Robert S. Kaplan. 1991. *The Design of Cost Management Systems: Text, Cases, and Readings.* New York: Prentice-Hall.

Cooper, Robin, Robert S. Kaplan, Lawrence S. Maisel, Eileen Morrissey, and Ronald M. Oehm. 1992. "From ABC to ABM." *Management Accounting* (November) pp. 54-57.

Cost Accounting Standards. 1962, 1987. In Chapter 1-1-3, *Rules of Business Accounting in Japan*, 20th ed. (first published in 1962). The Finance Ministry's Enterprise Accounting Deliberation Council: Dobunkan Shuppan.

Dilts, David M., and W. Russell. 1985. "Accounting for the Factory of the Future." *Management Accounting* (April) p. 37.

Drury, Colin. "Activity-Based Costing." 1989. *Management Accounting* (September) p. 66.

Hirabayashi, Yoshihiro. 1989. "Problems with Japanese Cost Accounting Standards." In *Japanese Management Accounting,* Y. Monden and M. Sakurai, eds. Portland, Oregon: Productivity Press.

Hiraoka, Hidehuku. 1994. "Reengineering and ABC." *Proceedings of Japan Management Accounting Association* (June) pp. 1-19.

Horváth, Péter, and Reinhold Mayer. 1989. "Prozeßkostenrechnung: Der Neue Weg zu mehr Kostenrechnung und Wirtungsvolleren Unternehmensstrategien." *Controlling* (Germany). (1. Jg. S) pp. 214-219.

Howell, Robert A., James D. Brown, Stephen R. Soucy, Allen H. Seed III. 1987. *Management Accounting in the New Manufacturing Environment.* Joint Project by NAA and CAM-1, National Association of Accountants. According to the U.S. survey, 62 percent of companies were dissatisfied (p. 41); in environments where machines were central, 24 percent of companies had adopted this and this was the greatest number (p. 131).

Hunt, Rick, Linda Garret and C. Mike Merz. 1985. "Direct Labor Cost Not Always Relevant at H-P." *Management Accounting* (February) p. 61.

Itoh, Hiroshi.1978. "Review of the Cost Accounting Standard." *Journal of Cost Accounting Research* (November) p. 33.

Johnson, H. Thomas. 1992. "It's Time to Stop Overselling Activity-Based Concepts." *Management Accounting* (September) p. 31.

Kagono, T., I. Nonaka, K. Sakakibara, and A. Okumura. 1985. *Strategic Versus Evolutional Management: A U.S.-Japanese Comparison of Strategy and Organization.* North-Holland.

Kaplan, Robert S. 1988. "One Cost System Isn't Enough." *Harvard Business Review* (January/February) pp. 61-65.

Kaplan, Robert S. 1990a. First Speaker of "Contribution Margin Analysis: No Longer Relevant/Strategic Cost Management: The New Paradigm." *Journal of Management Accounting Research* (Fall) pp. 3-5.

Kaplan, Robert S. 1990b. "The Four-Stage Model of Cost Systems Design." *Management Accounting* (February) pp. 22, 24-25, 26.

Kaplan, Robert S. 1992. "In Defense of Activity-Based Cost Management." *Management Accounting* (November) p. 59.

Kaplan, Robert S. 1994. "Management Accounting (1989-1994): Development of New Practice and Theory." *Proceedings of Edinburgh Department of Accounting/CIMA Joint 75th Anniversary Conference* (June 3) p. 8.

Keating, Patric J., and Stephen F. Jablonsky. 1990. *Changing Roles of Financial Management: Getting Close to the Business.* Morristown, New Jersey: Financial Executive Research Foundation.

Kobayashi, Kengo. 1978. "Review of the Cost Accounting Standard." *Journal of Cost Accounting Research* (December) p. 29.

Monden, Yasuhiro. 1993. "A Comparison of Japanese Kaizen Costing and Activity-Based Costing." *Accountant's Course* (October) p. 6.

Morrow, Mike, and Gray Ashworth. 1994. "An Evolving Framework for Activity-Based Approaches." *Management Accounting* (London) (February) p.32.

NAA (IMA) Tokyo Chapter. 1988. "Managerial Accounting in the New Manufacturing Environment: What We Learn from the Restructuring of American Managerial Accounting." NAA Toyko Chapter (October) p. 129. Those companies replying that improvement is necessary were 48 percent. This was roughly equivalent (50 percent) to those replying that the current cost system is reasonable. There was less dissatisfaction than in the United States. See p. 91 of the same work for methods of improving product costing.

Norkiewicz, Angela. 1994. "Nine Steps to Implementing ABC." *Management Accounting* (April) p. 30.

O'guin, Michael C. 1991. *The Complete Guide to Activity-based Costing*. New York: Prentice-Hall.

Pine, B. Joseph, Bart Victor, and Andrew C. Boyton. 1993. "Making Mass Customization Work." *Harvard Business Review* (September-October) pp. 108-119.

Raffish, Norm, and Peter B.B. Turney. 1991. "Glossary of Activity-based Management." *Journal of Cost Management* (fall) pp. 53, 55, 58.

Sakurai, Michiharu, and Ito, Kazunori. 1989. "Type of Industry and Managerial Accounting Practice." *Business Review of Senshu University* (October) p. 62.

Sakurai, Michiharu. 1986. "How Does FA Change Management Planning and Control Systems?" *Diamond Harvard Business* (March) p. 67.

Sakurai, Michiharu. 1994. "ABC: Overhead Management and Reengineering." *Management* Vol. 21 (May) pp. 79-83.

Scarbrough, D. P., A. Nanni, and M. Sakurai. 1991, 2. "Japanese Management Accounting Practices and the Effects of Assembly and Process Automation." *Management Accounting Research* (London) pp. 27-46.

Sharmen, Paul. 1990. "A Practical Look at Activity-Based Costing." *CMA Magazine* (February) p. 8.

Sonoda, Heizabrou. 1979. "How Cost Accounting Should Be Done." *Journal of Cost Accounting Research* (May) pp. 1-2.

Takayanagi, Kouichi. 1994. "ABM: A New Management Tool for Operations Management, Not to Make Reengineering Failure." *Nikkei Information Strategy* (July) pp. 135-144.

Tokai, Mikio. 1983. "The New Revolution in Manufacturing and Cost Accounting Systems." *Journal of Business Practices* (July) p. 24.

Turney, Peter B. B., and Allen J. Stratton. 1992. "Using ABC to Support Continuous Improvement." *Management Accounting* (September) p. 47.

Turney, Peter B.B. 1992. "Activity-Based Management." *Management Accounting* (January) pp. 20, 21.

Whitmore, John. 1906. "Factory Accounting as Applied to Machine Shops." *Journal of Accountancy* Vol. 2, no. 4 (August) pp. 248-258.

Yoshikawa, Takeo. 1994. *ABC Management for Restructuring/Reengineering*. Chuokeizai-sya.

CHAPTER 6

Measuring and Evaluating Quality Costs

"Made in Japan," once the object of scorn, has become a mark of distinction. The Japanese are the acknowledged leaders of the modern worldwide quality revolution, with brands like Toyota, Nikon, Sony, and Panasonic now virtually synonymous with high quality and reliability in the international arena. Yet, the Japanese do not view themselves as the winners of any quality competition, because they understand that the work of improvement is constant and grueling. In the business boom of the 1980s, Japanese businesses made quality improvements without careful cost-benefit computations. In the tougher current market, they are beginning to be more cost conscious about quality improvement. As this push continues in the high variety/low volume manufacturing environment, CIM has emerged as the most cost-effective way to improve quality, yet it also brings its own problems.

Morse and Poston (1987) have rightly stated that "the development and implementation of better quality cost measures and reporting practices should be an integral part of a CIM plan." In this chapter we will outline and discuss quality costing, clarify what it means, and show how to use it. We will also contrast the American practices of quality improvement with those in Japan.

Quality Cost Management in the U.S. and Japan

Maintaining and improving quality requires conscious improvement activities. Even in Japan it requires a constant effort and a sometimes difficult internal dialogue to maintain the focus on quality. Since the early 1980s, many American organizations have used Total Quality Management (TQM) to provide this focus, although their approach differs somewhat from that of the Japanese.

The U.S. accounting press poses the question, "Is there a trade-off between prevention costs and failure costs in achieving quality, and, if so, what is the optimal point?" Most quality related articles in *Management Accounting* and other core accounting journals center on this issue. Many U.S. companies try to measure and evaluate the cost of quality by studying the relationship between quality and cost structure as a part of their budgeting process. This area has received much less attention in Japan, leading to the question, "Why aren't Japanese companies interested in quality costing when the quality of Japanese goods is so highly appreciated?"

There are essentially three reasons. First, Japan is not plagued with the problem of poor quality, because of the success of its quality revolution. Therefore, there may be less need for the measurement of quality cost. (In the same way that recession begat cost accounting, poor quality may beget quality costing.) Second, rather than considering that there is a trade-off between cost and quality, Japanese managers typically aim for the highest quality. Third, little research is being done in the field of quality costs, except for the work of a few academics such as Kijima (1989), Sakurai (1990), Itoh (1992), and Murata, et al. (1995), even though there has been active research on quality in operational management areas. As a result, there is little discussion in the accounting literature, and financial executives are not aware of these concepts and their application to industry. Additionally, quality control in the United States focuses on testing in contrast to the Japanese view that quality products can be produced through TQC in the manufacturing process. This means that quality control research in Japan is likely to be very different from that in the United States.

What Is Quality?

Any discussion of quality costing must include a definition of what constitutes quality. There are three major interpretations in current use: degree of conformance, fitness for use, and innate excellence.

Degree of Conformance

In this view, quality is achieved if a product conforms with its specifications. However, this definition leaves some areas unaddressed. For example, even if a car manufactured by Toyota satisfies the conformance goals, it may not achieve the quality goal for riding comfort expected by the consumers.

Fitness for Use

This view is more user-oriented as it requires that the product meet the expectations of the customer (Garvin 1984). When quality is thus determined relative to the expectation of the customer, it is important to assess how well the company can satisfy the customer.

Innate Excellence

The third view is that quality is an innate characteristic of the essential superiority of a product or service. Under this definition, a high quality product must have appeal that will not change as time passes, regardless of how styles and tastes change. In contrast to the second meaning of quality this view is an absolute or universal one. For example, there are some people who would not think the work of Hiroshige (Ukiyoe painter) would complement their house decor, but few would deny the superior quality of his work.

American versus Japanese Perceptions of Quality

Generally speaking, American researchers consider quality cost to be the cost of conformance. For example, after defining quality, Roth and Morse (1983) state: "the quality cost discussed here deals with costs associated with quality of conformance as opposed to costs associated with quality of design." Morse, Roth, and Poston (1987) and Carr and Ponemon (1994) also focus on the costs associated with conformance activities. It is certainly easier for accountants to measure conformance costs than the other types of quality cost, and they do enable managers to focus on defective units, which has value.

In contrast, Japanese managers place more importance on "market quality," which means the difference between market or customer needs and product design specifications. In Japan, this type of quality is believed to be largely determined at the development stage (Itoh 1992), in much the same way that 80 percent or more of product costs are now thought to be determined in the design stage. Managing quality cost at the development stage is therefore the crucial quality activity, although quality control activities at the production stage are also indispensable. In Japan, "market quality" is seen to be multidimensional, and to include: conformity with specifications, appropriateness for use, functional performance, brand name, reliability, durability, maintainability, safety, and ease of use. Even in the United States, ideas of what constitute quality costs have been changing rapidly. For example, Dale and Plunket (1992) states that quality costs are the costs incurred in the design, implementation, operation, and maintnance of an organization's quality system, the cost of organizational resources committed to the process of continuous quality improvement, and the costs of system, product, and service failures. This multidimensional market quality concept is the basis for discussing quality in this book.

The Goal of Quality Costing

The purpose of quality costing is to produce a product with high quality at the lowest possible cost. It attempts to achieve this mainly by measuring the costs to the company of conformance failures. In the United States, where quality costing is carried out on a large scale, quality costs were once 10 to 20 percent of sales (although the active implementation of TQM has brought significant improvements). This compares to 2.5 to 4 percent in Japan (Roth and Morse 1983). The main cost drivers were the costs of defective parts, defective goods, the large amount of testing, repairs, and warranty costs. Reducing this ratio to 2.5 percent of sales has been the most important goal of American quality costing.

There are three major goals when implementing quality costing. The first is to know the nature and size of quality costs, which can be detected primarily through quality costing (Tyson 1987). This makes managers aware of the problems and gives them reasons for engaging in continuous improvements or reengineering. Second, quality reports that are aligned with both departmental and corporate performance evaluations provide management with an opportunity to take corrective action and are therefore useful in improving performance. Third, quality costing can improve the company's profitability through more effective budgetary control (failure cost, which is a cost driver, can become a budget line item), and more effective marketing activities.

As for the lower use of quality costing by Japanese managers, perhaps the generally high quality of Japanese manufactured products means that large quality improvements are not expected, and therefore the increased cost of accounting for the quality costs may be larger than the anticipated savings. However, as Kijima (1989) has pointed out, there is still ample room for quality improvement, even in the advanced technology fields. And the short life cycle of products in the current market means that even if you are proud of the present quality of your product you must still seek ever higher quality. The improvement of quality along with the reduction in costs is a never-ending pursuit and neither American nor Japanese companies can afford to ignore these issues.

What Is Quality Cost?

Quality cost can be narrowly defined as the cost incurred because poor quality may exist or does exist (Morse, Roth and Poston 1987). From this point of view, quality cost is the cost of doing things wrong. In some companies, these costs are very much a part of the culture. In the pre-AT&T-breakup Southwestern Bell Telephone Company, a common expression among the line workers was, "We may not have time to do it right, but we always have time to do it twice."

In a broader sense, quality cost includes three basic categories: costs incurred because of anticipated failures; costs incurred because failures were made; and costs incurred to create an environment in which employees can do their work effectively. These types of quality costs are not always available through the accounting system.

In the United States, quality costing generally measures the cost of the first two categories: 1) Costs incurred because poor quality of conformance *can* exist (prevention and appraisal costs), and 2) costs incurred because poor quality of conformance *does* exist (failure costs). These costs have four constituent elements:

- Prevention costs—Costs that are incurred in order to prevent the provision of products or services of inferior quality, such as education, training, and quality circle activities.
- Appraisal costs—Costs of inspections and tests to guarantee that products match specifications so that no additional work will have to be done.

- Internal failure costs—Costs due to defects or failures that occur prior to delivery of services or shipment of products to customers.
- External failure costs—Costs such as returned goods, discounts, and guarantees that must be given because defective goods were delivered or shipped to customers.

Of these, the prevention and appraisal costs are *voluntary* costs that are incurred by—and can be controlled by—management decisions. In contrast, the internal and external failure costs are *non-voluntary* costs that are incurred as the result of failures. It is important to separate them because the cost per failure is dramatically different for internally detected and externally detected failures, often 500 percent greater or more. Figure 6-1 shows the relationship between voluntary and failure (non-voluntary) costs, and the stage in which they are typically incurred.

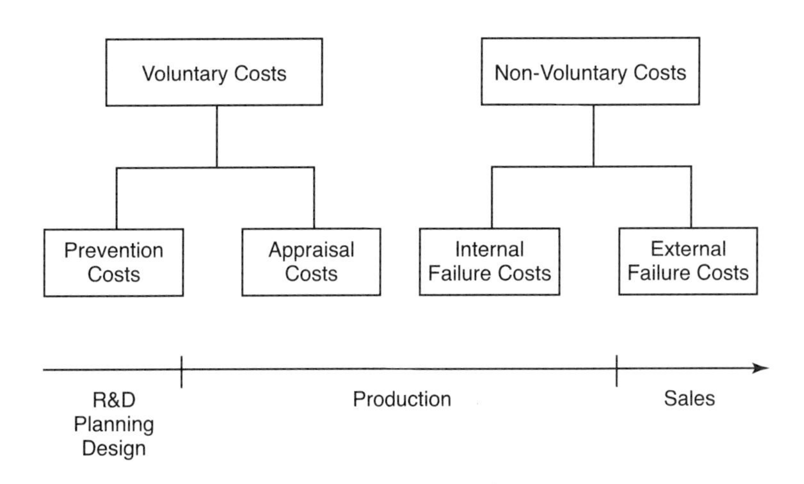

Figure 6-1. The Relationship between Voluntary and Non-Voluntary Costs

Prevention Costs

Prevention costs are incurred to prevent defects. From the financial perspective, they are more of an investment than an expense, even though US GAAP (Generally Accepted Accounting Principles) treats most of them as expenses for external disclosure. They are, so to speak, investments for future cost avoidance.

Typical prevention costs include: 1) the costs required to build a system for quality engineering, 2) the cost of promoting quality circles, 3) costs for education and training related to quality and work, 4) costs for preventing reoccurrence of failure (generally an engineering cost), and 5) costs for supervision and preventive maintenance.

Most prevention costs are managed costs. The typical Japanese manager believes that the most effective strategy for improving poor quality is to invest in prevention programs. However, even in Japan these programs are often neglected because measuring the effect of investments in prevention is difficult.

Appraisal Costs

Appraisal costs, also referred to as inspection costs, are the result of an inspection process in which outputs are evaluated to determine whether or not activities are being conducted properly (i.e., according to established standards and procedures). Typical appraisal costs include: 1) costs required for the tests performed by the company's quality control (QC) department, 2) costs of tests conducted by external research facilities, 3) costs required to guarantee quality in the manufacturing process, 4) costs for maintaining and calibrating test equipment, 5) costs of inspections for on-the-spot decision making, 6) cost for proofing and finalizing documents and packaging, and 7) costs for handling and reporting data on quality.

Some of the appraisal costs are variable, based on the amount of testing administered and the number of defective items identified. However, other costs are discretionary fixed costs, and are under the manager's control. The need for appraisal costs indicates a lack of confidence in the company's failure prevention activities. On the other hand, if TQC activities are performed thoroughly and successfully in a company it is theoretically possible to reduce appraisal costs to zero, and it is realistically possible to reduce them dramatically.

Internal Failure Costs

Internal failure costs are incurred due to failures detected by the company before products are passed on to the consumers. In other words, these are costs for eliminating failures found by inspections. It includes the costs incurred from the time materials and parts are shipped by suppliers to the time that the finished goods are received by the end users.

The following are some examples of internal failure costs: 1) the cost of failures in the manufacturing process such as scrap, spoilage, and rework, 2) the loss due to downgrades (inferior quality items are typically downgraded and sold at a lower price), 3) the costs for discovering breakdowns and repairing them, 4) costs for engineering changes because of poor quality, 5) the cost of computer reruns (re-execution of jobs), 6) the cost of any safety stock required for supporting low yields in processing, defective parts, or rejected lots, 7) the cost of reinspecting and retesting after a defect has been discovered, and 8) the loss due to work interruption.

External Failure Costs

External failure costs are incurred when defects in a product or service are found by external customers. These costs occur because the appraisal system failed to detect all the defects prior to shipping. External failure costs include: 1) the cost of returned goods, 2) the cost for granting a discount due to inferior quality, 3) the cost of lawsuits over defective items, 4) the administrative costs of handling claims, and 4) the costs of product recalls.

These four types of quality cost can usually be obtained from the accounting system without redesigning the entire system. They constitute what are often called *direct* quality costs. However, *indirect* quality costs cannot be measured by accounting systems and so are often not captured in quality costing. One of the drawbacks of traditional, functionally-oriented accounting systems is that the cost of activities is not usually revealed. For example, the costs associated with quality activities are disbursed across functional areas in most traditional accounting and budgeting systems. Each manager is responsible for only a portion of the cost.

A significant part of the logic behind quality costing is that if managers see the accumulated costs of poor quality, they will then have the will and ability to do something about it. Of course, it is not necessary to have accounting reports to realize this. Some managers probably have this perspective based on experience or aptitude, and it is probably easier to understand in smaller companies with more comprehensible relations between activities and outcomes. The popularity of quality costing in the United States is also consistent with the command and control philosophy so popular there, since it makes it easy to envision trading off these costs to achieve a global minimum.

Balancing Prevention/Appraisal Costs against Failure Costs

Prevention costs are all costs expended to prevent failures, or, to put it another way, all the costs involved in helping the employees do the job right every time (Harrington 1985). As Figure 6-2 shows, it is expected that as preventative activities proliferate, failure costs decrease due to the decrease in the number of failures. The curve on the right side tapers off because the management provides employees with the education, tools, equipment, systems, and knowledge to perform their work correctly.

In contrast, while appraisal activities reduce the chance of shipping defective products, strict testing alone will not decrease the occurrence of defective products, just their transmission to the outside world. While it is remotely possible that a strict testing regime will lead employees to improved quality (if the quality failures could be influenced by their efforts), it is much more

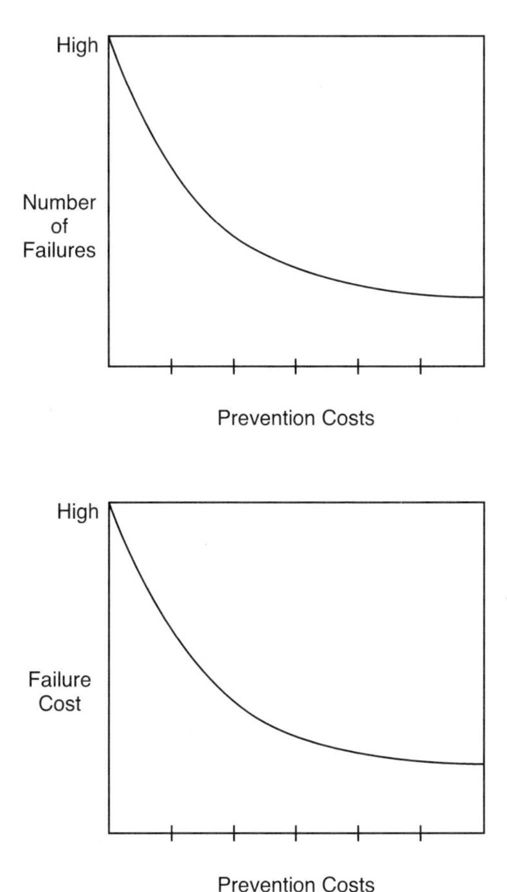

Figure 6-2. The Influence of Prevention Cost on Number of Failures and Failure Cost

likely that such a strict regime would lead to employee resistance. For example, at a jet engine turbine blade factory in the southeastern United States, such a regime lead the line employees to perform their own ad hoc quality inspections—hiding any defective parts under the floor boards of the old factory building. When discovered, the cache of blades numbered in the tens of thousands and represented a material portion of the factory's output.

Investment in preventative activities, on the other hand, usually has a higher return. A manager at Mitsubishi Electric gave this example: If a defective resistor is disposed of before use, it costs ¥60. If discovered at the time of assembling a circuit board, it would cost ¥2,000. If it were discovered to be defective only after it was in the hands of the client, it would cost millions of yen (more than ten thousand U.S. dollars.)

What then, is the relationship between voluntary costs and failure costs? Figure 6-3 presents a theoretical conception of these relationships. On the left side of the curve, when the voluntary quality costs are very low (no effort is made to avoid or detect failures before delivering the product to the customers), failure costs are very high. As voluntary costs are increased (due to quality improvement efforts), failure costs will decrease because the number of failures detected before delivering to the customer is decreased. On the right side of the curve, the voluntary costs increase conspicuously. However, because the increased efforts gradually become less effective, the decrease in cost becomes very small. For example, take the problem of checking a chapter in a book for mistakes. If the first proofreading revealed nine mistakes, the second proofreading might reveal only three mistakes. In the fifth proofreading, only one mistake might be found, yet the cost and time devoted to the fifth reading would be almost the same as the first.

When the voluntary costs and failure costs are added, a new curve emerges. An effective quality system is represented in this diagram by the low cost point on the curve. This is often called the "best operating point" in the United States. At this point, the sum of the voluntary quality costs and the failure quality costs become minimal and the ROI may be the maximum. "Best interim operating point" is probably more apt, however, because the point of lowest cost changes as process improvements lower the level of failure. Thus, the curve displays the relationship between voluntary costs and failure costs under a very specific production method, which in a IKM operating regime is very short lived.

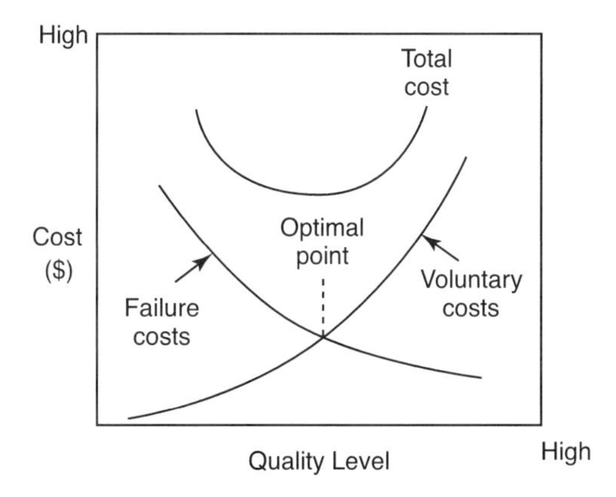

Figure 6-3. Theoretical Relationship between Voluntary and Failure Costs in a Fixed Production Environment

As a decision aid, this mechanism has weaknesses similar to those in economic order quantity (EOQ) analysis in traditional inventory management. Both models are equilibrium solutions, and as such they disguise the malleability of the entire cost structure in the interests of mathematical tractability. In theory, of course, this should not pose a problem since users are assumed to be able to correctly interpret the impact of the underlying computational assumptions, however, most accounting academics now realize that this is wishful thinking.

A second and possibly equally important problem is that the goals on the revenue side are not addressed. The model could be extended to include the revenue side by introducing the concept of quality elasticity of demand. Then the profit maximizing point on the line would not necessarily be the lowest cost point. In fact, it is an implicit assumption embedded in most U.S. quality costing literature that consumers are quality neutral. This model illustrates the fact that lowest cost will not be best unless it is also the benefit maximizing point.

Very few Japanese believe in the optimal point shown in Figure 6-3 in quality costing. They believe it is necessary to make failure cost zero by investing more in voluntary costs. What is the actual proportional relationship between voluntary costs and failure costs? Harrington (1987) suggests that for the typical commercial product in the United States it would be 45/55. Recent research (Atkinson et al. 1994) suggests no big difference from the earlier estimate. At the recent meeting on quality costs many Japanese participants agree that the ratio would be 60-70/20-40 depending on how prevention costs are defined (see Table 6-1).

Table 6-1. Distribution of Quality Costs in the U.S. and Japan

Types of Quality Costs	U.S. Companies		Japanese Companies	
Voluntary costs		45%		60–70%
Prevention cost	10%		40–50%	
Appraisal cost	35%		20%	
Failure costs		55%		20–40%
Internal failure cost	48%		–	
External failure cost	7%		–	
Total quality costs		100%		100%

Compared to Japanese management practice, prevention costs in the U.S. are quite small. American managers have typically relied on inspections, not TQC, to maintain quality levels (Gray 1981), so that for Americans, quality control normally meant testing or appraisal activities. In contrast, Japanese managers believe that quality can only be improved by continuous improvement participated in by all members of the company. Therefore, they are very active in TQC, which requires all departments and all workers to be involved in quality improvement. This understanding is now taking hold in many of the better American firms as American businesses turn their attention to quality improvement. The development of TQM in American companies is one explicit example (GAO 1991, George and Weimerskirch 1993).

Use of Quality Costing

When quality costs are measured, how are the results used in practice? One of the ways to put quality cost to use is through ratio analysis.

Ratio Analysis of Quality Costs

Performance analysis begins with totaling quality costs. While it is important to know the amount spent on quality for each product, common denominators must be devised in order to facilitate comparisons between factories or divisions. The quality costs of one factory or a division may be greater than another simply because of size differences. In such circumstances, a common denominator is necessary in order to make a fair comparison.

The choice of an appropriate common denominator for a company will depend on its organizational structure. According to Edmonds, Tsay and Lin (1989), when the division is organized as a *profit center*, one could use the following three indexes as common denominators: total sales, unit sales price, and net profit. In contrast, when the division and factory are organized as *cost centers*, the common denominator may be expressed as a proportion of assets, of direct labor hours, or of manufacturing costs.

Edmonds et al. (1989) provide the following three indexes as examples used in the quality cost manual of the Formosa Plastic Group (FPG): 1) the average quality cost per unit of sales (quality cost divided by sales amount), 2) the average quality cost per manufacturing cost (quality cost divided by manufacturing cost), and 3) the average quality cost per direct labor hour (quality cost divided by direct labor hour).

These ratios are three logical indexes that a company might use in attempting to present quality costs in a way amenable to management action. The effectiveness of using ratio analysis at the division level depends on whether a useful common denominator can be found, and on whether the company can overcome all of the normal problems associated with any performance measure.

Trend analysis for quality costs can be successfully conducted by using bar charts, line graphs, or pie charts to show ratio information in a more useful way. For example, Omron, a top Japanese manufacturer of control devices, uses the Q cost ratio. This is calculated as quality cost divided by production costs (factory invoice price to sales department). This ratio is reported to top management once a month using bar charts.

Management at Omron found Q cost to be a useful common denominator, since direct materials costs, direct labor cost, and factory overhead are already expressed as a proportion of Q cost. The cases of FPG and Omron provide examples in which quality cost can and should be used in performance evaluation.

Budgeting for Quality Costs

In addition to its use for performance evaluation, top management can view each department as a responsibility center for quality improvement with a goal that can be expressed in the budget or the profit plan. Goals that are frequently used by companies include the reduction of: voluntary costs, failure costs, prevention costs, appraisal costs, internal failure costs, and external failure costs. When sales are buffeted by stiff competition in an industry, managers will probably turn their attention to external failure costs. Normally, however, managers are concerned about the effect of internal failure costs on manufacturing efficiency. When upper-level managers want to set a positive goal for controlling management activity, prevention costs and appraisal costs are good budget targets. Additionally, when anything becomes a line item on a budget it gets more separate attention.

Because accomplishing quality goals takes a long time, special attention should be given when preparing a budget for quality cost. Traditionally, budgeting was directed toward short term cost achievement goals. For example, a manager would be responsible for producing a given amount in a period of time with planned direct material, direct labor, and factory overhead costs. When the actual cost is less than the standard cost, the manager would be evaluated highly. On the other hand, when the actual cost exceeds the standard cost, managers responsible for the work would get a low rating.

If this type of plan were applied to controling quality costs, the results could be counterproductive because it would tend to reward quality *reducing* behaviors. To lower quality costs in the short run, a manager might be tempted to decrease expenditures on prevention and appraisal activities, or to ship sub-standard products to customers to avoid internal failure costs. However, if this happens repeatedly, external failure costs will ultimately increase. The result will be dissatisfied customers and higher quality costs in the long run. In this scenario, lowering quality costs in the short run may harm the image of the product and the company, and eventually the bottom-line. Consequently it is necessary to consider very carefully how to include quality costs in the budgeting process so that it is coherent with the long-term operating strategy.

It may even be necessary to change managers' basic long-term strategy in order to incorporate quality costs in the budgeting process. For example, the Japanese managerial philosophy takes a longer view than the short-term position generally taken by Americans. At the very least, American companies, while having some sort of short-term goal, must operate from a long-term perspective with respect to quality cost.

Reporting Quality Cost

Both American and Japanese companies prefer statistical methods to accounting systems for measuring defects, returned goods, and customer claims. A 1987 telephone survey of American companies revealed that at that time 31 percent of corporate controllers were measuring quality costs on a regular basis (29 of the 94 Fortune 500 companies surveyed) (Tyson 1987). What sort of system do they use to measure quality costs? In a survey by Howell, Brown, Soucy and Seed (1987), 50 percent of the responding companies measured quality cost. Of these companies, only 10 percent did so through the accounting system, 11 percent through the operating system, and 29 percent measure quality costs only when necessary. More than one-third (39 percent) of the respondents did not measure quality costs at all.

A similar study conducted in Japan reported that about one-third (32 percent) of surveyed Japanese companies practice quality costing (Sakurai 1992). It may be a surprise to the readers that such a large number of Japanese companies actually practice quality costing because it was generally believed that quality costing is rarely practiced in Japan. However, only 8 percent have quality costing integrated with budgeting and responsibility accounting. The remaining 24 percent use it only as a tool for engineering. The same study also reported that more than half (56 percent) of those companies practice TQC but not quality costing.

Table 6-2 is an example of a quality cost report used by Japanese companies. Internal and external failure costs are not separately totaled here (even though they can be spotted easily enough), but other companies do separate external failure costs from internal failure costs. In this example you can see that quality cost is compared with manufacturing cost. The manager in the quality assurance department considers the trends in prevention cost, appraisal cost, internal failure costs, and external failure costs to be the most important aspect of this report.

This type of report is generally prepared by the quality assurance department in both American and Japanese companies. For example, at Omron, this report is prepared jointly by the quality assurance department and the engineering department, with no accounting assistance.

Notes for Using Quality Costing

There are several things to be aware of when putting quality costing to use. The following are three essential considerations.

Table 6-2. Quality Cost Report

Description	Current Month		Year to Date	
	Quality Cost	% of Mfg Cost	Quality Cost	% of Mfg Cost
Prevention: (Target)				
Quality planning Quality engineering Quality training Data review Quality diagnosis Total cost				
Appraisal: (Target)				
Product inspection Reliability test Tool inspection, etc. Total cost				
Failure: (Target)				
Scrap Rework Claims Replacement Warranty Total cost				
Total quaity cost				
Total quality target				

Long term perspective

The results of quality improvement efforts will not show up quickly. Therefore, as Morse and Poston (1987) pointed out, it is important to take the time lag between the efforts and the results into consideration, especially when reporting quality cost.

Quantifiable improvements are only part of the quality improvement result—an important portion of the improvements in quality are difficult to quantify. Over-emphasizing the measurable results of quality costing may tend to encourage a short-term orientation in operations. It is important to maintain a long-term perspecitve when applying the results of quality costing.

Subjective judgment

Determining what are and what are not quality costs calls for subjective judgments. Take prevention costs as an example. In a factory where paint is applied, a product may be finished with one coat of paint. However, with two coats there is not only a greater gloss but a decrease in defects, and the product is more durable. The cost for the second coat of paint thus improves quality, but is it a quality cost? There is no general answer; however, the criterion must be consistently applied.

Indirect quality costs

Indirect costs normally cannot be obtained from the company's accounting system. Indirect costs are defined as opportunity costs and life-cycle costs (which will be discussed in Chapter 7). There are three classes of indirect quality costs.

1. *Costs incurred on the user's side*, such as time lost due to equipment breakdowns, decrease in the productivity of equipment, repair costs incurred after the expiration of warranty, clerical costs incurred because defective products are returned, and so on.
2. *Costs arising from customer dissatisfaction.* It appears that there is no quality cost in accounting terms when a product matches the appropriate quality level. However, when there is a gap between product design specifications and customer expectations then there will be a quality cost. Customers' expectations for quality have grown markedly in Japan. If companies merely interpret quality as meeting design specifications, their product quality may quickly become inferior in relation to the market.
3. *Effects on the company's reputation.* A poisoning incident at Morinaga & Co. (a general confectioner with a long history

in Japan) due to lack of attention during manufacturing directly affected not only the company's profits but also its reputation. No one knows the long term effect on profitability; however, its image is profoundly tarnished.

Despite the fact that they cannot be quantified, attention must be given to these opportunity costs. When performing quality costing, it is necessary to devise a system that will call attention to these types of indirect quality costs. Another way of looking at these last two indirect costs is to regard them as part of the revenue side of the quality issue, since opportunity costs are usually unavailable revenues.

The Practice of Quality Costing in Japan

How is quality costing being practiced in Japan? The following eight cases gathered from company visits will illustrate the practices and characteristics of Japanese quality costing.

Company A—A World Class Automobile Manufacturer

During the plant visit, a manager in charge of target costing said that 100 percent quality must be insured in automobile production. Therefore target costing is conducted in the design stages, based on the assumption that quality is always perfect without any allowance for failure costs. Thus, prevention and appraisal costs must be thoroughly realized and actions should be taken to bring internal and external failure costs close to zero. Yet, defective work does occur in the manufacturing process. Although quality costing provides management with quality costs, it cannot itself improve quality. This is because accounting figures are useful primarily for optimal deployment decisions, not change decisions. Other operational systems are required for quality improvement. That is the main reason why TQC is essential for minimizing failure costs in their company.

Company B—A World Class Computer Company

In the past, Company B's computers were "absolutely guaranteed not to break down," a controller there told us. In order to promote sales in the late 1960s and early 1970s, they sold computers to the Central Racetrack Authority on the condition that Company B would be responsible for all costs when a breakdown occurred. This is not an unusual practice in Japan. In horse racing, computer breakdowns simply cannot be permitted because a crucial part of the value is the immediate payoff for winning bets. Therefore, at each stage of the manufacturing process, Company B created checkpoints to thoroughly inspect the quality of the product by imagining the existence of as many failures as possible and then trying to follow them up. All members of the company have made use of the experience of that time to enable them to produce products with high quality and 100 percent reliability.

One of the characteristics of Company B is that they operated on the premise of keeping failure costs, internal or external, down to zero. In terms of their quality cost structure, in the beginning they operated with high appraisal costs because of the many checkpoints, but now they are able to operate with no failure costs relying primarily on prevention costs.

Company C—A Leading Manufacturer of Electrical Products

Company C keeps track of only internal failure costs because of the difficulty of measuring prevention costs. However, they do not use failure costs for management control purposes. Instead, they use the raw data without converting it into monetary amounts because they believe that improving quality first hand has the greatest impact on the workplace. In other words, they are practicing TQM rather than quality costing. The basic approach of Company C is to manage quality by linking it with TQC. It is Company C's belief that quality is "created" in the process of target costing. That is, when the target costing committee deals with a new product they give the highest priority to quality, delivery time, and cost. What they emphasize is cost reduction with highest quality.

Company D—A Pharmaceutical Company

At Company D, the removal of impurities is essential to the very survival of the company. Quality is therefore paramount at Company D. Consequently, prevention and appraisal costs are everything and no failure cost is allowed. It is true that there are some claims, but regardless of the magnitude of the claims, their policy is not to trade off the claims cost against the prevention cost. Using Company D's definition of prevention cost, it is as much as nine percent of the manufacturing cost.

The Ford Motor Company got into serious trouble when it was revealed that they had explicitly made this computation when evaluating the cost of relocating the gas tank to a safer position on the Ford Pinto. Basically, they estimated that the cost of any additional wrongful death lawsuits would be less than the cost of the engineering change. Although this is felt by many to be a rational way to make business decisions, the damage to Ford's reputation in some circles was significant.

Company E—A Plastic Fabricating Sub-Contractor

Quality is also of absolute importance at Company E, a plastic fabricating sub-contractor with annual sales of ¥10 billion (US $100 million). Upon the request of the parent company (and main customer), managers at Company E report failure costs. Since they produce products to meet the specifications required by the parent company they have no use for the full range of quality costs. However, in the future, the required specifications are expected to become even more demanding. If these cause higher costs, it may be necessary for the company to install more sophisticated quality costing.

Company F—A Telecommunication Company

Company F, the biggest telecommunications company in Japan, always focuses on worker and customer safety in planning and performing its constant repair activities. It has been measuring safety costs and the cost of safety education, though not quality costs, for years. Their information shows that safety education has raised the quality of service. Since the quality of service is related to customer satisfaction (Hatayama 1988), safety costs at Company F may resemble the prevention costs in manufacturing industries.

Company F introduced quality costing to equipment maintenance in 1992. There are three types of quality costs: 1) preventive costs for equipment improvement and maintenance; 2) operation costs for supervision and maintenance activities; and 3) failure repair costs. There is a trade-off among the three costs in this industry.

The underlying force for installing quality costing is the recent severe competition in the changing business environment. Because of budgetary constraints there is a tendency for managers in this company to decrease expenditures on preventive activities. As a result, equipment has become more worn out and the number of breakdowns has increased remarkably. This is a clear example of a drop in quality leading to the use of quality costing. The main purpose of installing quality costing was to show how much damage the company would suffer if prevention expenditures are reduced to below a sufficient level.

Company G—A Construction Company

Company G built an embankment to guarantee safety though it did not measure quality costs. The company has been successfully measuring safety costs for many years. Whether one can include these safety costs as integrated construction costs and charge them to the client depends on the agreement between the client and contractor.

Company H—A Maker of Control Equipment

Company H is a manufacturer of control equipment. Quality costing was started in 1989 when the Managing Director of the company realized the need for reducing failure costs. Quality costs in this company are categorized into preventive, appraisal, and failure costs. The company does not divide failure costs into internal and external failure costs because the identification of failure cost is more important than their classification. Quality of conformance may be the goal of quality costing in major American companies, but fitness for use or customer expectation is much more important for Company H. Thus, the company does not use the concept of the quality of conformance. The trade-off between cost and quality, therefore, has never been under consideration in this company.

Quality costs are integrated with budgeting but not related with the cost accounting system. The results of quality costing are used for both the improvement of operations and performance evaluation in business units. The former is mainly used by workers on the shop floor and the latter by the top management. The company keeps track of three types of quality measurements: quality cost, Q cost ratio (Q cost/total shipment cost), and failure cost ratio (failure cost/total shipment cost). Quality costs are measured and reported once a month. As of August 1991, claims from customers and failures in production processes decreased by 7 percent and 21 percent respectively since 1990. As of March 1992, the Q cost ratio was 6.2 percent in 1990, 5.02 percent in 1991, 4.2 percent in 1992, and 3.4 percent in 1993. This relationship is shown in Figure 6-4.

The Characteristics of Quality Costing in Japan

Few Japanese companies would explicitly trade quality for cost. Quality is not considered as a means to achieve the goal of cost reduction. Japanese management tends to think that both quality and cost reduction should be the goals of the company, even if costs may sometimes rise in pursuit of the best quality.

In the 1960s, the following expression was used frequently, "A mistake will be allowed once, but not a second time." However, according to Mr. Jinichi Kamiki, the chief of the Quality Control

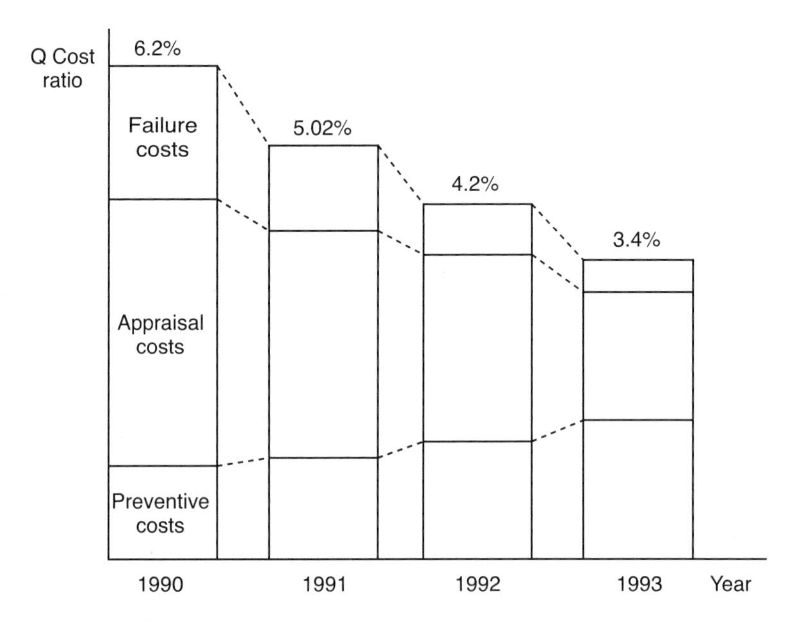

Figure 6-4. Effect of Q Cost System in Company H

Department at Snow Brand Milk Products, nowadays mistakes cannot be permitted at all (Kamiki 1990). The severe competition of the modern market place will not permit failures in planning, design, and production. Companies must guarantee that the planned product will appear on the market at the planned time, with the planned quality. This means that the consumers have become very demanding, quality has improved greatly, and competition among companies has become extremely intense.

If the purpose of management is to know the trade-off between quality and costs, then the accounting system is needed to report quality costs. However, since one of the chief goals of Japanese firms is to improve quality (not short term divisional performance) quality costing is normally conducted by engineers who are not involved in accounting. One of the most desirable approaches to quality costing is first to plan a high quality level and reasonable cost in setting the target cost, and then to do everything possible to improve quality in the manufacturing stages

—through QC circles or other continuous improvement methodologies.

Thus, quality costing in Japan, presuming that it does exist, does not attempt to measure or manage quality cost in order to determine a trade-off between quality and cost. Instead, it is seen as a tool for quality management which supports the production of high quality products throughout the planning, design, and manufacturing processes.

Conclusion

Both the goals and methods of quality costing differ between American and Japanese companies. First, quality control in the United States is usually conducted by the quality control department, and it is generally applied during the production stage. This can be verified by the fact that prevention costs amounted to only 10 percent of the average U.S. total quality cost, as shown in Table 6-1. In Japanese companies, on the other hand, quality control means TQC, and quality cost is mainly composed of prevention costs that are expended throughout the production process. Since preventive efforts to improve quality occur simultaneously with production processes, it is difficult to quantify prevention costs. However, they would likely amount to between 40 and 50 percent of the quality costs in a Japanese company. This is a fundamental difference between the Japanese and American approach to management, although some changes are taking place. Intense foreign competition in general, and Japanese competition in particular, has led some American companies to adopt TQM practices that are prevention based. This approach is sometimes referred to as "building-in quality." Juran (1993) argues that American businesses will regain quality leadership, rung by rung, when American senior managers carry out the quality managment roles they cannot afford to delegate. For details, see United States General Accounting Office (GAO 1991).

The second major difference relates to which costs are measured, Normally, the costs measured for quality costing are limited to those direct quality costs that can be quantified by the accounting system. Probably because of their tractability they are the main object of quality costing in the western world. However, even in the United States there are several academics who talk about the importance of indirect quality costs. For example, Kaplan and Atkinson (1988) maintain that omitted from the calculation are the costs of disruption in operations caused by out-of-conformance purchases and production, and the loss of sales caused by actual external failures and associated reputational effects. However, quite a few academics write about quality costing based only on quantifiable figures. In contrast, almost every Japanese manager advocates including those costs that cannot actually be obtained from the accounting system. One of these quality costs is the loss of reliability that damages a company's image and its future bottom-line when defective units are sold to customers. The effects of this type of loss last for a long time and inflict considerable damage to a company. Though these losses are difficult to quantify, they are among the most important quality costs.

Improved quality can be one of the goals of the conversion to CIM, although improvements in quality are also very difficult to quantify. Nonetheless, the reduction in quality cost from quality improvement should never be ignored in operating strategy. Issues relating to CIM investment will be discussed in Chapter 8.

Thirdly, American and Japanese businesses differ in their understanding of the relationship of quality and cost. Writers in the United States maintain that quality costs should be measured and reported as part of the process for improving quality. Nonetheless, the relationship of quality and cost is generally taken as a trade-off between the two in the United States (Nakane 1990). The idea that costs should be kept down and that quality should be improved only up to a given point is well-entrenched in American business. This is because the user or consumer will prefer the inexpensive alternative if they can get a given level of quality.

In contrast, the general tendency of the Japanese consumer is to try to get the highest level of quality possible even if the products are low-priced commodities. This is because Japanese users have a higher expectation on quality and will not accept defective or inferior goods. The "Survey of American and Japanese Prices" conducted in December 1990 by Nikkei (1991) validated this difference. They found that with reference to shopping behavior, 26.4 percent of the Japanese replied that they "would feel more comfortable buying a more expensive item if there was a slight difference in price between two items," while only 9.6 percent of the Americans surveyed provided the same answer. Whereas 44.4 percent of the Japanese replied that "they would not want to purchase a product made by an unknown manufacturer even if it were cheaper," only 7 percent of the Americans voiced the same opinion. Though the price of beef in Japan was three times higher than that in the United States at the time of the survey, 44.4 percent of the Japanese replied that "when shopping for food they would buy the expensive and fresh items," whereas only 13.5 percent of the Americans replied this way. 52.8 percent of the Japanese and 15.4 percent of the Americans replied that "it was more economical to buy higher quality clothes even if more expensive since they could be worn longer." It is clear from these results that the Japanese consumers are more conscious about quality than price. The survey was conducted on 100 representative households in Tokyo and New York. The average age of the respondents was 41 years and the households had 2 children. The average annual income for the Japanese household was ¥8,580,000 ($65,000) and for the American household, $70,075 (¥9,250,000) assuming an exchange rate of 1:132. 72 Japanese and 52 American households responded.

It is clear that Japanese managers are less apt to emphasize the trade-off between quality and cost. The fact that they have been able to produce high quality products may be attributed to their consumer's high expectations. Yet, we doubt that it will be possible to continue to seek the highest level of quality while ignoring cost. More Japanese companies will need to measure quality cost with special cost studies, and it is likely that the technique they will use will be target costing.

References and Further Readings

Atkinson, Handy, John Hamburg, and Christopher Ittner. 1994. *Linking Quality to Profits: Quality-Based Cost Management.* ASQC Press, pp. 13-26.

Carr, Lawrence P., and Lawrence A. Ponemon. 1994. "The Behavior of Quality Costs: Clarifying the Confusion." *Journal of Cost Management* (summer) pp. 27-29.

Dale, Barrie G., and James J. Plunkett. 1992. *Quality Costing.* Chapman & Hall, p. 5.

Garvin, David A. 1984. "Product Quality: An Important Strategic Weapon." *Business Horizons* (March- April) pp. 40-41.

GAO (United States General Accounting Office), Report to the Honorable Donald Ritter, House of Representatives, Management Practices. *U. S. Companies Improve Performance Through Quality Efforts.* (May 2) p. 2.

George, Stephen, and Arnold Weimerskirch. 1993. *Total Quality Management: Strategies and Techniques Proven at Today's Most Successful Companies.* John Wiley & Sons, pp. 1-269.

Gray, C. S. 1981. "Japan: Quality Control & Innovation—Total Quality Control in Japan." *Business Week* (July 20) pp. 23-24.

Edmonds, Thomas P., Bor-Yi Tsay, and Wen-Wei Lin. 1989. "Analyzing Quality Costs." *Management Accounting* (November) pp. 27-28.

Harrington, H. James. 1987. *The Improvement Process: How American Leading Companies Improve Quality.* McGraw-Hill Book Company, p. 46.

Hatayama, Yoshio. 1988. *What is Quality of Service?* Japan Management Association, p. 31.

Howell, Robert A., James D. Brown, Stephen R. Soucy, and Allen H. Seed III. 1987. *Management Accounting in the New Manufacturing Environment.* NAA, pp. 58-59, 39.

Itoh, Yoshio. 1992. "Practice of Quality Costing." *Accounting* Vol. 44 no. 8, pp. 32-40.

Juran, Joseph M. 1993. "Made in U.S.A.: A Renaissance in Quality." *Harvard Business Review* (July-August) p. 48.

Kamiki, Juichi. 1990. "Fundamental Approach in Setting Policy."

Survey of Management Issues Production, Japan Management Association (November) p. 42.

Kaplan, Robert S., and Anthony A. Atkinson. 1989. *Advanced Management Accounting*. Prentice-Hall, p. 382.

Kijima, Yoshitaka. 1989. "The Concept of Quality Costing." *Accounting* Vol. 41, no.11 (November) p. 86.

Morse, Wayne J., and Key M. Poston. 1987. "Accounting for Quality Costs in CIM." *Journal of Cost Management* (fall) pp. 5, 9.

Morse, Wayne J., Harold P. Roth, and Key M. Poston. 1987. *Measuring, Planning, and Controlling Quality Costs*. National Association of Accountants, pp. 10-12, 2.

Murata, Naoki, Noriyoshi Takeda, and Keiich Numa. 1995. *Discussion on Quality Costing—Its Origin and Development*. Taga Shuppan Ltd., pp. 1-290.

Nakane, Jinichiro. 1990. "Will American Manufacturing be Revived? The Present State and Trends in Production Systems in American Manufacturing Companies." *Factory Management* Vol. 36, no. 1(January 1990) p. 70. Though the same professor says "where they are seriously thinking of trying to revive manufacturing, the trade-off idea is in retreat" (p. 71).

The Nikkei. 1991. "Survey of American and Japanese Prices" Nihon Keizai Shinbunsha Inc. (March 3) pp. 28-29.

Roth, Harold P., and Wayne J. Morse. 1983. "Let's Help Measure and Report Quality Costs." *Management Accounting* (August) p. 50.

Sakurai, Michiharu. 1990. "Measurement and Appraisal of Quality Costs." *Journal of Business Practice* no. 436, Enterprise Management Association, IMA Japan Chapter (August) p. 17-27.

Sakurai, Michiharu. 1992. "Japanese Management Accounting Practices: Analysis of Mail Survey for CIM." *Business Review of Senshu University* no.55 (October) p. 155.

Tyson, Thomas N. 1987. "Quality & Profitability." *Management Accounting* (November) p. 39 (According to Tyson's survey of 29 representatives of quality control departments, 24 indicated that they are using quality cost, 23 for evaluating performance, and 10 for budgetary purposes.)

CHAPTER 7

Life Cycle Costing

Life cycle costing is a method used for calculating the costs of products or equipment over their entire life. This method is used for various management purposes, such as for capital budgeting decisions, or when trying to produce quality products at a lower total life cost. These costs fall into two broad categories: manufacturing costs and user costs. On the manufacturing side, life cycle costs include all the costs that the producer will incur over the product's life cycle. On the user side, it includes all the costs that the user will incur to obtain, use, and dispose of the asset.

Though companies show strong interest in the manufacturing costs (in the case of manufacturing companies) or purchasing costs (in the case of user companies), they have not always consciously considered the costs incurred after acquiring the assets. This is primarily because the costs of operation, maintenance, and disposal are difficult to measure, and the costs themselves have been relatively low. However, this is changing. Recently, the market for such high technology products as computers, aircraft, and computer software has increased significantly. For the business planner, a key feature of these products is that they obligate the owner to extraordinarily large operation, maintainence, and/or disposal costs. These post-purchase costs will often be many times greater than the purchase cost. For decisions about such products, there is an increasing need to measure and analyze total life cycle costs, including not only manufacturing or purchase costs and marketing costs, but also user costs for operation, maintenance, and disposal.

The Concept Of Life Cycle Costing

In traditional cost accounting, the life cycle of a product or asset includes R&D, planning, design, and manufacturing. Of these, all of the costs that incur through planning, design, and manufacture also appear in traditional product costing. R&D costs are accounted for either as product costs or period costs, and all these costs incur under the responsibility of the manufacturer. However, the actual life cycle does not end when the product has been manufactured. There is some confusion here because the term *product life cycle* is used in two different ways in the literature. Under one definition, it refers to the state of production, sales, and earnings from the time a given product is placed on the market until sales are discontinued. The more comprehensive view refers to use of a product from the time it is manufactured or purchased until it is discarded. According to Imai (1980), the term *life cycle* generally refers to the former, but it is used in the latter sense in reference to durable goods in Japan. Life cycle profit management as suggested by Susman (1989) addresses mainly the former sense (the periods of introduction, growth, maturity, and decline). In contrast, in life cycle costing, the latter meaning is generally used since the costs from R&D and product planning to disposal are the objects of its study. It has also become very important for companies to consider life cycle management issues for computer software (see Lehner 1989; Fabrycky and Blanchard 1991).

The traditional market life cycle activities are shown on the left in Figure 7-1. These activities generally are carried out when the manufacturer is responsible. The broader definition of life cycle also includes the sustaining activities shown on the right side of the figure. These include operation, maintenance, and disposal costs. The actual life cycle ends when the product or equipment can no longer be useful or is worn out. The activities from operation to disposal generally fall under the user's responsibility.

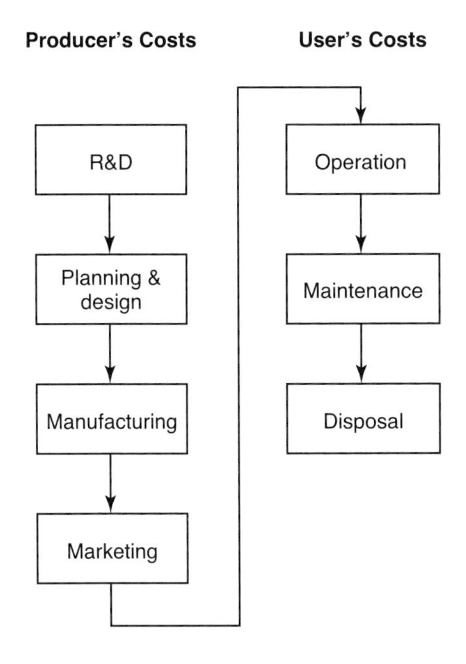

Figure 7-1. Actual Life Cycle of a Product

The Significance of Life Cycle Costing

Life cycle costing has significance both for manufacturers and users. Manufacturers traditionally show interest in the manufacturing costs that occur up until the time the product is transferred to the user. They have not explicitly shown much concern over the costs that accrue to users after the products or equipment have been transferred to them. However, the intensified competition of today's market along with the advance of high technology means that the responsibility of the manufacturer no longer ends with producing a product that matches stated features and specifications. To be competitive today, a manufacturer must design a product from the start that improves quality, reliability, and support in order to optimize performance and profitability for the user.

The user of an asset measures all the important life cycle costs incurred during the lifetime of an asset, typically measuring life cycle costs with a cash flow analysis as part of a capital assets evaluation model. In the model, cash flow is discounted to its present value so that the user can have a means of making an optimal selection of assets. This allows the user to consider the trade-offs between cost elements during the asset life phases. For example, a user may choose a higher initial cost in order to secure a future reduction in maintenance costs.

Life cycle costing is also necessary when making decisions concerning operation and maintenance costs incurred during the life of the asset. While life cycle cost is usually considered at the planning and design stages of a product or piece of equipment, some argue for a change in focus. The increased interest in disposal costs resulting from the advance of technological innovation and the related shorter product life suggests the need for a move from a design-to-cost focus at the design stage to life cycle analysis at the R&D stage (Takemori 1990).

Cost-benefit analysis, cash-flow discounting, sensitivity analysis, probability theory, and other techniques for capital budgeting are all applied in life cycle costing. These are not new concepts at all, in fact, they are actually nothing more than a combination of techniques to make managers think about the cost of managing assets in order to maximize the asset's value. The wide range of life cycle costing literature (Gupta and Chow 1985) suggests there are many issues to be discussed in life cycle costing.

Product Life Cycle and Life Cycle Costing

When talking about life cycle costing, some writers discuss issues in the product life cycle in addition to changing user costs. For example, Susman (1989) and Czyzewski et al. (1992) include discussions of the product life cycle when they talk about life cycle costing. By the same token, Berliner and Brimson (1989) describe marketing, engineering, logistic support, and discontinuation of a product as the chief elements of life cycle cost issues. The Japan CPA Association (JICPA 1985) views the life cycle as having three elements: initial costs (R&D costs, planning, design, etc.), normal costs (manufacturing costs, sales costs, etc.), and final costs (repair costs, discontinuation costs, etc.).

The Relationship with Design-to-Cost

Life cycle costing has been practiced for a long time in the Defense Department of the United States. It is applied to all new weapons systems in proposal or during development. Since its impact has been very great in defense industries and in space exploration, these industries tend to design products based on life cycle costing. This practice is referred to as "design-to-cost." Design-to-cost establishes the desired cost (life cycle cost to the maximum extent feasible) as a design parameter during a system's design and development phases, and provides a cost discipline to be used throughout the acquisition of a system (Department of the Army, the Navy, and the Air Force 1977).

Design-to-cost, like life cycle cost, includes the costs of development, production, operation, support and, when applicable, disposal (Department of the Army, the Navy, and the Air Force

1977). In other words, the range of the costs in question, and the range of life cycle costing, will in part overlap. In that respect life cycle costing and design-to-cost are understood to mean the same thing.

Some researchers argue that design-to-cost is similar to target costing in design/planning/execution because design proceeds after establishing a target price (Ezaki 1984). Certainly, design-to-cost is similar to target costing in that it is a cost management technique applied in the product planning stage. However, as Makido (1986) has pointed out, in design-to-cost, the target cost is initially estimated from engineering knowledge, which differs from the Japanese target costing practice in which the target cost is determined from the sales price less desired profit. Thus, target costing in the Japanese manner is tightly and strictly tied to market demands. In this crucial respect design-to-cost is different from target costing.

The Method of Life Cycle Costing

The most effective strategy for reducing a product's total life cycle costs is to focus cost reduction efforts on those activities that occur before manufacturing begins. Thus, it is the best policy for manufacturing companies to analyze life cycle cost at the time target costing is applied.

The Steps of Life Cycle Costing

The process of life cycle costing differs between manufacturing and user companies. In life cycle costing and target costing, both manufacturer and user life cycle costs should be considered. For manufacturing companies, life cycle costing is typically applied at the planning and design stages. In these stages, Japanese companies attempt to "build cost and quality into" the product. As Artto (1994) rightly argues, manufacturers' life cycle cost analysis can provide valuable cost information for target costing. The goal of life cycle costing is to take actions and decisions that cause a product to be planned, designed, marketed, distributed, operated, maintained, and disposed of in a way that promotes the long-term competitive advantage of the company.

For user companies, life cycle costing is generally a three step process for capital asset acquisition. The first step is to clarify the need for the asset in question based on consideration of the business environment and the goals of the company.

The second step is to purchase an asset that matches those needs at the lowest possible lifetime cost. To make this decision the company needs:

- to predict the requirements for the asset and the cost of the asset
- to quantify the life cycle cost of alternative assets
- to decide upon the best proposal

After the purchase, the third step of life cycle costing is to compare and analyze the actual cost over the lifetime of the asset with the target cost.

The Cost Categories in Life Cycle Costing

The second step in life cycle costing is the analysis of trade-offs between the initial cost of acquisition or purchase and the subsequent costs that arise in connection with operation, maintenance, and disposal. Life cycle costs may be roughly divided into three groups.

- the costs related to initial capital investment
- the costs incurred in operating and maintaining the asset for use
- the disposal cost

Initial capital investment

The following costs are included in the initial investment for companies that design and build physical assets for use or resale.

- R&D
- Design and specifications
- Manufacturing/construction
- Installation
- Writing of manuals and training

In financial accounting these costs are often measured as assets. Generally, this sort of data is easily obtained from accounting records.

There is some question about whether R&D costs can be capitalized. In the United States, R&D costs should be expended (FASB number 2). In Japan, however, they can either be expended or capitalized (deferred assets) selectively in financial accounting (Commerical Law 286, Article 3). According to Berliner and Brimson (1988), automation of factories has transferred the arena of competition to new product development and new process development, making substantial investment in R&D necessary. Because these investments will influence the cost structure of a company over the long run, they should be treated as capital investment and attributeed to the cost of the product, rather than as periodic costs for management accounting. Many Japanese companies include applied and industrial R&D costs in product costs, while basic R&D is expended.

Operation and maintenance costs

The second group of costs are incurred in the operation or maintenance of an asset. The following are included in these types of costs.

- The cost incurred in operating an asset (including labor costs, materials costs, costs for tools, and other overhead)
- The cost incurred in maintaining an asset (including the cost of supplies and labor)

The above include operating and maintenance costs for computer software as well as physical equipment. The following might also be included in equipment operating costs.

- Lost opportunities from production due to breakdowns or breakdown in maintenance
- Low utilization arising from the fact that equipment does not match needs, or the product or service is not considered necessary
- Decline in performance that occurs because the equipment cannot produce output at the level of quality required

Companies need to measure such lost-opportunity costs in life cycle costing. Unfortunately, they are very difficult to obtain from the financial accounting structure, so typical Japanese companies consider these costs but do not include them in the financial accounting mechanism.

Disposal costs

Finally, there is the salvage value and disposal cost. An automobile that you have driven for five or six years has a salvage value when you get rid of it. On the other hand, disposing of nuclear fuels is very expensive. In order to properly evaluate these costs, it is necessary to estimate both the salvage value and disposal cost. The disposal cost is usually the easier of the two to estimate. Due to today's social and environmental factors, disposal of equipment is becoming more and more expensive.

Trade-Offs in Life Cycle Costs

The importance of trade-offs is highlighted by the reciprocal relationships among the three categories of cost: initial capital investment, operating and maintenance costs, and disposal costs. For example, if a lot of insulation is used when building a structure, the amount of initial capital investment will be greater but operating costs will be lower in the future.

Figure 7-2 shows a well-known diagram for balancing optimal life cycle cost among the cost factors of initial capital investment, operating and maintenance costs, and disposal costs (Taylor

1981). This diagram suggests that though in the first option, initial capital investment is small compared to operating and maintenance costs, in the second option, by doubling the amount initially invested, operating and maintenance cost were significantly reduced.

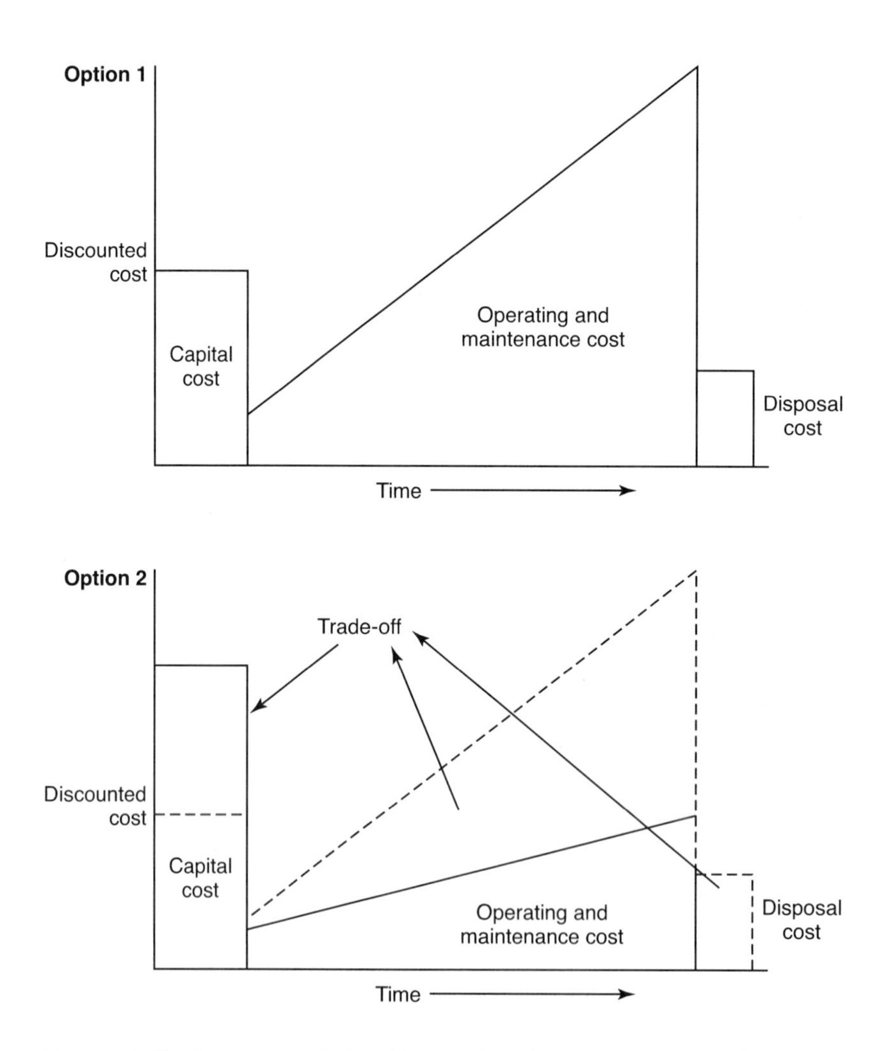

Figure 7-2. The Cost Factor Relationship over the Lifetime of a Capital Investment

Figure 7-3 concerns the popular relationship between equipment and purchasing and shows how the life cycle cost is optimized. As the initial amount of capital invested in equipment represented by the slope of curve A increases, the cost of operation and maintenance represented by the slope of curve B decreases. Curve C represents the total of initial capital investment and operating and maintenance cost. The optimal life cycle cost is the lowest point on curve C.

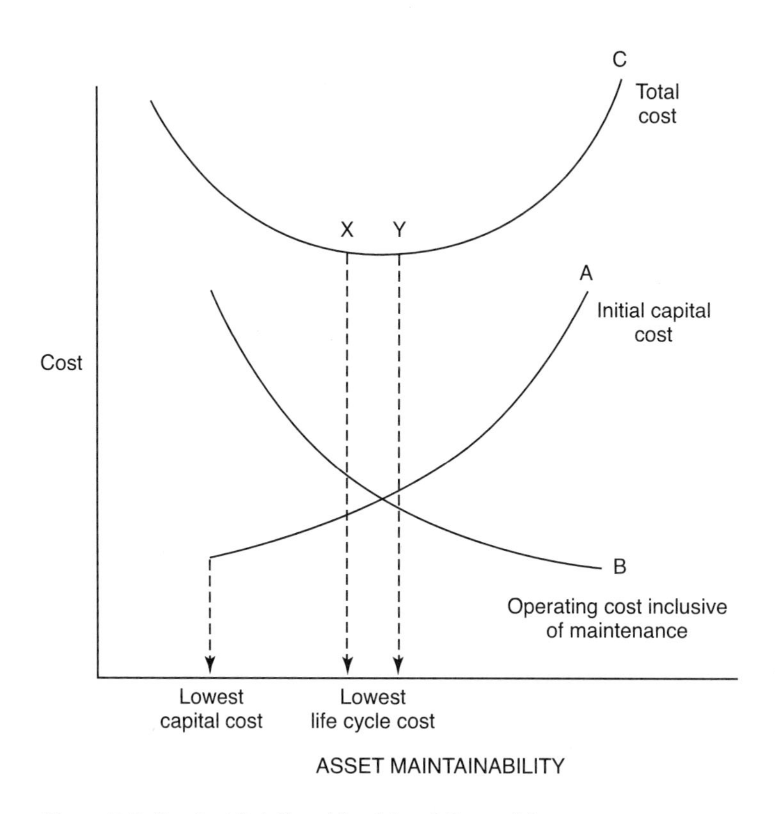

Figure 7-3. The Cost Relationship of Asset Ownership

Americans tend to focus on the optimal selection of costs from existing options in their discussions of life cycle costing. There is little discussion of reducing the size of the life cycle cost by reducing the cost associated with some or all of the three phases. It is always assumed that there is a reciprocal relationship between these costs, and that the effective manager has navigated the best route down the total cost curve. What is often forgotten is that these costs, being estimates, do not exist yet, and are not out there waiting to be discovered, but are awaiting creation. The Japanese IKM approach, on the other hand, does not rest on such assumptions, but views costs as plastic, even when borrowing life cycle costing tools from the United States.

The Application of Life Cycle Costing

Western life cycle costing approaches presuppose a desire to trade-off between the initial investment, and operating, maintaining, and disposal costs. This can only be effectively applied in some areas and industries.

Areas Where Life Cycle Costing Can Be Effectively Applied

There are five major factors that affect the economic feasibility of applying life cycle costing.

Operating and maintenance costs

Life cycle costing is most effective when significant savings can be gained by reducing maintenance costs. For example, when using computers, there are significant maintenance costs. According to one survey, the relative percentage of maintenance costs is 67 percent of all information processing costs (Bell and Pugh 1992). This large investment in downstream costs is the reason for life cycle costing when manufacturing or acquiring computer software.

High energy consumption

When energy consumption by equipment or an acquired product will be significant during its lifetime, life cycle costing is very effective. For example, the amount of energy consumed by aircraft, automobiles, ships, and boilers is large, so life cycle costing is especially pertinent in such industries.

Length of expected useful life

For equipment and products that will last a long time, costs incurred after acquisition become more significant than the initial amount invested. On the other hand, for equipment with a short usable life, the initial cost is more important. For example, though the legal lifetime of a helicopter was only two years in Japan (the law was changed to five years for those put into service for operations after April 1, 1990), they can be used for almost unlimited periods of time by replacing parts and components. Thus, life cycle costing is indispensable for helicopters. In contrast, the usable life of a Walkman is short, so application of life cycle costing is out of the question.

The amount of initial capital investment

Generally, when the amount of capital initially invested is large, that alone increases the importance of life cycle costing (Brown, et al. 1985). For example, since the initial capital investment in nuclear power generation equipment is very great, it is necessary to compute the cost over the entire life cycle in order to compare it with that of oil burning equipment.

Disposal cost (salvage value)

If the net disposal cost (salvage value less disposal cost) is significant, it is important to consider the life cycle cost. The most common reason for a high disposal cost today is the difficulty of finding environmentally sound disposal methods at a low cost. In a similar vein, when purchasing an automobile, individuals often consider the salvage value (the expected resale value).

Typical examples of capital assets for which all five of the above factors should be considered are: computers, electronic equipment, military equipment, automobiles, software, buildings and other facilities to be constructed, and equipment to prevent destruction of the environment. These industries not only require large initial investments but also incur significant operating and maintenance costs after acquisition.

Regaining Competitive Advantage with Life Cycle Management

Successful application of life cycle management can provide companies with competitive advantages. Companies that sell products of high value with superior quality satisfy their customers, and as a result, are able to dominate the market. As Figure 7-4 shows, the life cycle cost is related to cost, quality, and service, and ultimately is linked with customer satisfaction.

High quality products and superior service are needed to satisfy a customer's needs. High quality means a safe product with superior features, performance, and reliability. The product must be delivered on time, with good service and proper follow-up. When a product fulfills these requirements, and can provide not only a low manufacturing cost but also low operating, maintenance, and disposal costs, then it can beat the competition in the marketplace.

Successful application of life cycle management can yield customer satisfaction (Adamany and Gonsalves 1994). For example, costs for the use of a computer by the customer are often high, and the amount of profit that comes from service after the sale is almost the same as the profit that comes from sale of the product. Generally, product sales reach a temporary peak and subsequently decline but the profits obtained from maintenance continue. Companies that have loyal customers are able to reap substantial profits from the support, service, and maintenance departments. If they can design a more reliable product, they could maximize the earnings from maintenance. If they can improve quality, they can minimize claims from customers and reap the advantages of reduced parts in inventory and a higher level of customer satisfaction.

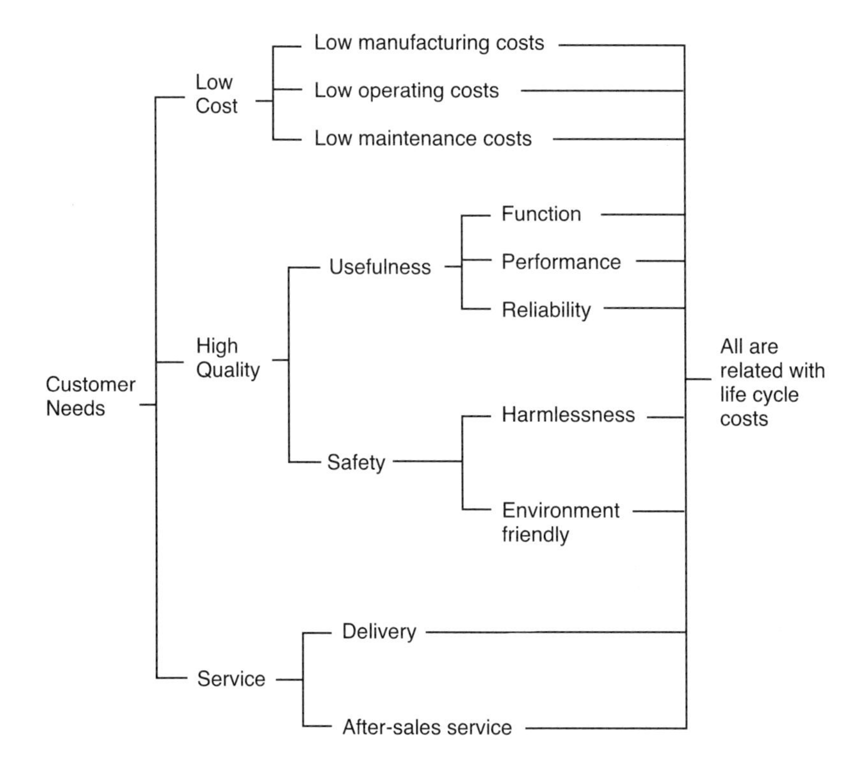

Figure 7-4. Cost, Quality, Service, and Life Cycle Costs

Recovery of American Automobile Manufacturing Industry

The recent recovery of the American automobile manufacturing industry is remarkable. When the author visited Ford Corporation in February 1993, the corporate controller of the company reported that Ford could beat all the Japanese automobile companies except Toyota. This success was the result of high quality, reliability, and the low price of American automobiles.

The turbulence of the past ten years in the automobile industry forced managers to recognize the impact of the complete product life cycle on costs. A number of American academics maintained that the American automobile industry fell significantly behind the Japanese automobile industry by ignoring problems that occur after manufacturing. Twenty years ago it was a common belief that American automobile manufacturers were concerned only with sales and did not focus on improving their products so they would last longer. The expression "planned obsolescence" characterized the manufacturers, who it appeared wanted to force consumers to buy cars more frequently.

From the point of view of life cycle management, quality costs can be broken down into manufacturer's quality cost, consumer's quality cost, and societal quality cost (Mizuno 1984). Under this breakdown quality costs for the automobile consumer would be the costs of fuel, inspection and maintenance, repair, and deterioration. The societal cost would be the burden to society calculated as the damage to the environment from exhaust gases. In other words, it is possible to take into account not just the manufacturing costs but the total costs to the consumer and to society by applying life cycle costing.

Life Cycle Management in Japan

Life cycle costing has had considerable theoretical development but few practical applications in the United States (Adamany and Gonsalves 1994). The same is also true in Japan. Life cycle costing is an expensive activity, so it does not ordinarily have to be applied both in the development of a new product and in the acquisition of all capital assets. The Japanese believe that life cycle costing should be applied only when the cost of its analysis is less than the possible benefits that will be gained from acquiring the equipment or product.

Life Cycle Income Statement

The result of life cycle costing can be effectively expressed in a Life Cycle Income Statement. Tables 7-1(A) and 7-1(B) present a

Life Cycle Income Statement as prepared by a Japanese manufacturing company. This company prepares both their income statement and the user's statement. They refer to the first as the Life Cycle Income Statement and the last as the User Cost Income Statement.

Table 7-1 (A). Life Cycle Income Statement

Sales		¥ 1,300,000
Less cost of the product		
R&D costs	100,000	
Planning & design	30,000	
Manufacturing costs	700,000	
Marketing costs	50,000	
Physical distribution	20,000	¥ 900,000
Income from the product		¥ 400,000
After sales costs		
Warranty costs	150,000	
Recall-related expenses	20,000	
Litigation expenses	10,000	180,000
Net income from the product		¥ 220,000

Table 7-1 (B). User Cost Income Statement

Purchase cost	¥ 1,300,000
Chargeable expenses	10,000
Operating costs	300,000
Maintenance costs	200,000
Disposal costs	90,000
Total user costs	¥ 1,900,000

The sales amount on the Life Cycle Income Statement in Table 7-1 (A) corresponds to the purchase cost on the User Cost Income Statement in Table 7-1 (B). The Life Cycle Income Statement is divided into two parts: costs needed to produce and market the product, and costs incurred after sales. The first is called cost of the product. Sales minus cost of the product equals income from the product. The second is called after-sales costs. Income from the product minus after-sales costs equals the net income from the product. Costs on the User Cost Income Statement consist of purchase costs, chargeable expenses, operating costs, maintenance costs, and disposal costs.

It is most effective to prepare the Life Cycle Income Statement at the planning and designing stages of a product. It is easier to reduce total costs without impairing product quality at this stage than trying to reduce costs during manufacturing. For example, the management of this company found that after-sales costs were very high. Fortunately, it was not too late to take effective measures to reduce after-sales costs. If the statement had been prepared after production activities had begun, the company would not have been able to reduce total costs. This is a good example of using life cycle costing to reduce costs rather than to trade them off.

The Status of Life Cycle Costing in Japanese Companies

The Japanese Defense Ministry and Japanese defense contractors have maintained a close relationship with the U.S. Department of Defense. As a result, when Japanese companies manufacture defense products for Japanese Defense Forces, they must follow the American life cycle costing practices. For example, the defense products manufacturing divisions at Kawasaki Heavy Industries, Mitsubishi Heavy Industries, and Ishikawajima Harima Heavy Industries have been using life cycle costing in the same manner as American companies for many years.

However, typical Japanese business practices differ from American practices in several ways. Life cycle management has mainly been practiced by engineers, not accountants (Okano 1994). As a result, life cycle costing as practiced by the typical American company has not taken hold in Japanese companies outside the defense industries.

Generally, manufacturing companies check life cycle costs at the product planning stage. For users, a purchasing decision often focuses on operating and maintenance costs. Typical Japanese users purchase products or equipment based on total life cycle costs, not simply initial purchase price or manufacturing cost. Life cycle management is already built in to many Japanese management practices, even though life cycle costing as a discipline had not gained much popularity. "Though no doubt there is increasing interest shown in life-cycle costing" as was pointed out by the Committee for the Japan Institute of Life Cycle Costing (1986), "the time is not yet ripe for really getting into Western-style life cycle costing."

Table 7-2. Degree of Satisfaction with Life Cycle Costing by Industry

	Producer	User	Type of users				
Questions			Engine	Chemical	Machinery	Steel	Metals
Interest in LCC	51%	51%	36%	50%	54%	61%	44%
Organization for LCC	19%	15%	8%	14%	19%	17%	18%
Use of LCC	22%	22%	17%	20%	24%	27%	20%
Level of satisfaction	29%	28%	20%	26%	31%	33%	26%
No. of plants	20	151	9	71	42	16	11

Table 7-2 shows the degree of satisfaction (by industry) with the use of life cycle costing. This survey was conducted by JIPM in 1983. Questionnaires were sent to its 523 member firms. Of these, 222 were engineering and machinery companies, of which 20 replied (9 percent response). The overall response rate was 29 percent. Although a comparatively low rate of satisfaction with life cycle costing was expressed, less than 30 percent of the users and less than 10 percent of the manufacturers responded. This may indicate that the level of interest in life cycle costing is still rather low in these industries.

The respondents identified two areas of problems in implementing Western style life cycle costing:

1. problems with the companies themselves
2. problems peculiar to life cycle costing

Within the companies, respondents felt they were not set up to accept life cycle costing. There was very little interest in, or understanding of, life cycle costing by management. There are also some problems peculiar to life cycle costing itself. For example, it is difficult to compute user costs such as operation, maintenance, and disposal costs. In addition to these difficulties, conducting life cycle costing would take much time and effort, evaluating the results would also take time—and there were certain aspects that would be hard to evaluate. In other words, the cost is high and the benefits uncertain.

The Importance of Life Cycle Management

What is practiced in Japanese companies today is not life cycle costing per se, but life cycle management. Life cycle management, as described below, attempts to manage and market/use a product or piece of equipment through the various stages of its life cycle. It includes total cost reduction and quality improvement programs for customer satisfaction.

Viewed from a historical perspective, cost management in Japan focused on manufacturing costs in the 1960s. In the 1970s it extended its reach upstream to include R&D costs and cost reduction activities at the planning/design stages, as well as marketing costs. Recently, especially after the 1980s with the increased volume of high technology products, the interest of Japanese managers is being directed toward user costs such as operating and maintenance costs. The importance of life cycle management can be seen in the following four points.

1. The focus of cost reduction in Japan has moved upstream as the opportunity for cost control in the manufacturing stage decreases due to factory automation. As a result, attention is directed to cost reduction activities in the R&D and product planning and design stages. For example, the costs of R&D continue to increase annually in Japan. This is illustrated in Figure 7-5. The companies shown here are all listed on the Tokyo Stock Exchange, first section, in the Nikkei database. As this figure indicates, the R&D cost-to-sales ratio increases very steeply each year.

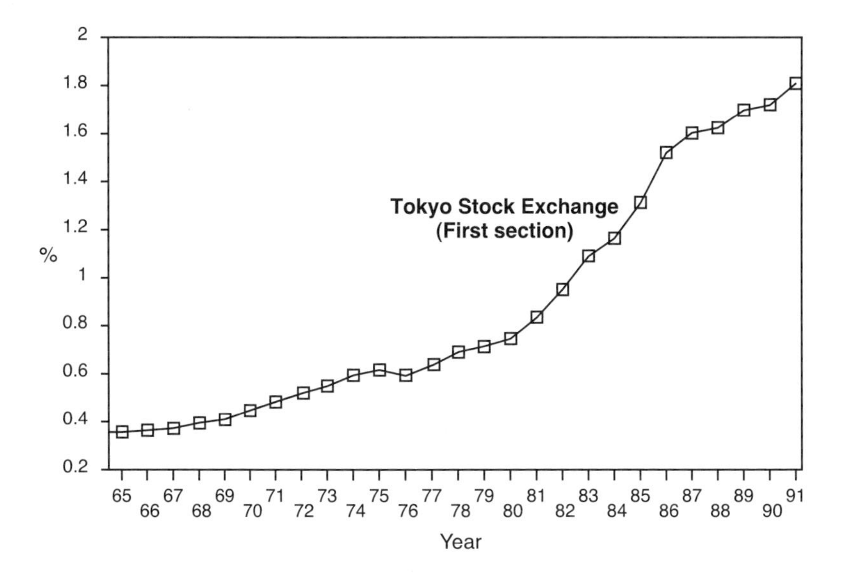

Figure 7-5. Ratio of R & D Cost to Sales

2. Downstream costs such as those for physical distribution have also increased in recent years. Therefore, Japanese managers have turned their focus to physical distribution costs in addition to manufacturing costs.

3. Together with the increase in the number of high technology products, the costs for operating and maintaining products has increased significantly. Purchasing CIM equipment is very costly; however, the costs incurred after equipment has been installed may be even larger. Thus, when CIM is installed, cost reduction activities for operation and maintenance have a decisive role.
4. The costs of disposing of products and equipment when a product is discontinued have also increased year by year.

Japanese Life Cycle Management Today

Recently, there has been an active application of life cycle management in the environmental equipment industry. In contrast to American practice, Japanese equipment manufacturers are customarily responsible for the equipment they sell even without any legally binding warranty. Manufacturers often perform maintenance service long after the sale. Life cycle costs in Japan also include environmental conservation equipment, and so place a much greater cost burden on companies than in the United States, South Korea, and other countries (Oshima 1990). In the United States and South Korea, the responsibility of the manufacturer usually ends with the sale, when responsibility is passed to the user. In the United States, this is often indicated by contract or is legally mandated. It is the same in South Korea and other Southeast Asian countries.

In contrast, Japanese manufacturers are bound by much long guarantee periods. Electronics products must be repaired by the manufacturer free of charge to the purchaser for one year, not for the ninety days common in the United States. Automobiles have a warranty for a given period (three or five years depending on general or special warranty) or a certain number of miles (60,000 or 100,000 km depending on general or special warranty) in all Japanese companies, compared with one year or twelve thousand miles in Ford. It has been standard practice that department stores replace commodities with even the slightest defect (this is also common in the United States). It is the implicit practice in Japan that stores and manufacturers cannot escape responsibility for the quality of a product they have manufactured or sold after it has been passed to the consumer, even without a contract. The costs for guaranteeing a product increase the required sales price of the product and weaken its competitiveness in the marketplace. These costs also create constant pressure to control costs and improve quality.

In the machinery industry, if a manufacturing machine breaks down and the user sustains a big loss, the user will not reorder from that manufacturer. So, with or without a contract, manufacturers must sell products and equipment that will last a long time.

In addition, Japanese manufacturers must think of conserving the environment and must install pollution prevention and control devices. For example, Nissan Automobiles substituted an alternate chemical for the ozone depleting foaming agent used in the production of its urethane bumpers in the middle of 1991, and in the middle of 1993 adopted a new cleaning agent for its parts. Fujitsu reduced its consumption of freon from 540 tons in 1986 and 600 tons in 1987 to 400 in 1989 and 260 in 1990. Though the increase in costs for reducing the use of freon cannot be predicted, Fujitsu plans to discontinue using freon altogether by 1995 (Diamond Weekly 1990).

Japanese companies such as Mitsubishi Heavy Industries and Ishikawajima Harima Heavy Industries boast the highest level of technology for pollution prevention in the world. "Japanese manufacturers through repeated trial and error have developed and produced technologies for stack gas desulphurization, the lime-gypsum method of removing sulphur oxide, and denitrating technologies, especially removing nitrogen oxide using ammonia (catalytic reduction method)" (Shibuya 1990). Though pollution prevention measures themselves raise manufacturing costs, Japanese companies bear these costs within the context of their responsibility to society and are approaching problems related to destruction of the environment positively.

Conclusion

In this chapter, we have discussed the concept of life cycle costing and its relationship to life cycle management. The Japanese approach to life cycle costing differs from the approach used in the United States in which there is an explicit focus on trade-offs between the cost to the user and the manufacturing cost. The Japanese have actively tried to cope with the life cycle, as with quality costing, in order to improve reliability during the period in which a product or piece of equipment is used. The Japanese appear to use life cycle management on a partly intuitive basis, rather than using life cycle costing as a discipline. In contrast to engineers, only a few Japanese accountants (for example, Someya 1976; JICPA 1985; Makido 1986; Sakurai 1990; Takemori 1990; Okano 1994; and others) have shown interest in life cycle costing.

The reason life cycle costing is popular in the United States but failed to take hold in Japan (although life cycle management appears to be popular), has much to do with cultural differences between the two countries, such as the following.

First, it is necessary for managers in both the United States and Japan to consider sales, technology, and production from the viewpoint of life cycle management. Japanese managers, in general, have endeavored to make reliable, high quality products. This has been more important to them than finding a cost-benefit trade-off between manufacturing cost and the user cost. Therefore, when Japanese managers evaluate Western-style life cycle costing with its explicitly analyzed trade-offs between cost and quality, Japanese managers fear that cost may be over-emphasized at the expense of producing a product with high reliability, which could lead to the loss of Japan's hard-won reputation for quality.

Second, the life cycle cost cannot be fully understood when the manufacturer and the user are different organizations. From the point of view of the manufacturer, apart from sample surveys, it is very difficult to know the user cost. Additionally, it takes a long time to accumulate the data needed to understand the cost to the user. If managers attempt to make a preliminary conclusion from only readily available data, there is a strong danger that management will be motivated by short-term considerations. Yet if we try to study this with a long-term view, there will be many cost items that will not fit into the categories of life cycle costs.

Third, it is preferable that an engineer should be in charge of life cycle costing. If accountants dominate the life cycle costing practice in a firm, they are likely to overemphasize the financial aspects. In the United States, life cycle costing is sometimes conducted as a financial process, as can be seen in capital budgeting. In this case management may become motivated by short-term considerations due to the financial emphasis. Thus, life cycle costing may be more effective as an engineering tool, with accountants serving to support the information needs of the engineers.

References and Further Readings

Adamany, Henry G., and Frank A.J. Gonsalves. 1994. "Life Cycle Management: An Integrated Approach to Managing Investments." *Journal of Cost Management* (summer) pp. 35, 39-40.

Artto, Karlos A. 1994. "Life Cycle Cost Concepts and Methodoligies." *Journal of Cost Management* (fall) pp. 28-32.

Bell, Doug, and John Pugh. 1992. *Software Engineering: A Programming Approach*, 2nd ed. Prentice-Hall, p.8.

Berliner, Callie, and James A. Brimson, (eds.). 1988. *Cost Management for Today's Advanced Manufacturing*. Harvard Business School Press, p. 140.

Brown, Robert T., and Rudolph R. Yanuck. 1985. *Introduction to Life Cycle Costing*. Prentice-Hall, Inc., p. 4.

Committee for the Japan Institute of Life Cycle Costing. 1986. *Report of Research Project Concerning the Life Cycle Cost of Maintenance Technology in Manufacturing Plants*. Japan Institute of Plant Maintenance (April) p. 199.

Czyzewski, Alan B., and Rita P. Hull. 1992. "Improving Profitability with Life Cycle Costing." In *Emerging Practices in Cost Management*, Barry J. Brinker, ed., pp. G2-1-G2-8.

Department of the Army, the Navy, and the Air Force. 1977. *Joint Design-To-Cost Guide, Life Cycle Cost As a Design Parameter*. AFLCP/ AFSCP 888-19 (April 28) pp. 1, 8.

Diamond Weekly Special Edition, 1990. "A Test of Companies Allegiance and the Makeshift Response of Large Companies: Is Environmental Preservation a Cost or a Benefit?" *The Diamond Weekly*. (June 16) pp. 17-18.

Ezaki, Michihiko. 1984. *New Thinking and Its Procedure for Design-To-Cost*. Sanno College Isehara, Publishing Department, pp. 2-4.

Fabrycky, Wolter J., and Benjamin S. Blanchard. 1991. *Life Cycle Cost and Economic Analysis*. Prentice Hall, pp.1-14.

Gupta, Yash, and Wing Sing Chow. 1985. "Twenty-Five Years of Life Cycle Costing—Theory and Applications: A Survey." *International Journal of Quality and Reliability Management* Vol. 2, no. 3, pp. 51-76.

Imai, Kingo. 1980. *Theory and Reality of Life Cycle.* Japan Management Association, p. 3-4.

JICPA (Japan Institute of Certified Public Accountants)/MS Consulting Office.1985. *Report of the Council on Common Problems, Part Two, Present Issues in Cost Control* (July) p. 54.

JIPM. 1983. *Report on the Present State of Life Cycle Costing in Japan.* Japan Institute of Plant Maintenance (September) pp. 8-9, 46-47.

Lehner, Franz. 1989. "The Software Life Cycle in Computer Applications." *Long Range Planning* Vol. 22, no. 5, p. 41.

Makido, Takao. 1986. "Life Cycle Costing." *Issues in Modern Cost Accounting,* Proceedings of the 45th Annual Meeting of the Japan Accounting Association, pp. 92-101.

Mizuno, Shigeru. 1984. *Companywide Quality Control.* Japan Science and Technology Federation, p. 172-174.

Okano, Kenji. 1994. "Introduction to Research on Life Cycle Costing. *Journal of Matsuyama University* (December) pp. 251-255.

Oshima, Masakatsu. 1990. "The Problems of Accounting in Developing Countries." *Management Review, Asia University Management Association,* Volume Commemorating the 20th Anniversary of the Founding of the Management Department at Asia University, Vol. 26, nos. 1 and 2 (November) pp. 65-66.

Sakurai, Michiharu. 1990. "Life Cycle Costing: Concept and Its Use." *JICPA Journal,* no. 424 (November) pp. 38-44.

Shibuya, Kazuhiro. 1990. "Report: Japanese Companies Which Cannot Make a Contribution to the 'World Environment or International Society.'" *Nikkei Business,* special supplementary edition (June 11) p. 184.

Someya, Kyojiro. 1976. "The New Appreciation for Life Cycle Costing." *The Waseda Commercial Review* no. 260 (November) pp. 1-17, 18.

Susman, Gerald I. 1989. "Product Life Cycle Management," *Journal of Cost Management* Vol. 3, no. 2 (summer) pp. 18-19.

Takemori, Kazumasa. 1990. "The Control of Research and Development Costs in LCCM." 49th Annual Meeting of the

Japan Accounting Research Association, Independent Papers 1, Session 4 (September 11) pp. 3-4.

Taylor, W.B. 1981. "The Use of Life Cycle Costing in Acquiring Physical Assets. *Long Range Planning*, Vol. 14, no. 6, pp. 37, 38.

CHAPTER 8

Investment Justification in CIM

Traditionally, any discussion about investment justification focused on whether the payback method or discounted cash flow (DCF) method is more appropriate. However, this is no longer the only issue when the investment decision relates to automated equipment such as FA and CIM. In contrast to traditional types of investments in plant and equipment, investment for automation does not simply reduce labor input. Many difficult to quantify indirect and intangible benefits result from such an investment. Any analysis of the investment decision that does not include these indirect and intangible benefits is incomplete.

Differences between Traditional and Automated Equipment

There are a number of differences between traditional facilities and equipment on the one hand and automated plants and equipment on the other. The following are some of the key differences.

Purpose

Investment in traditional equipment and facilities is intended to increase productivity by reducing labor costs and by conserving energy. However, automated equipment not only reduces labor but serves other purposes as well, such as improving quality, reducing inventories, reducing floor space, and providing freedom from the need to perform dangerous tasks.

Capability

Traditional equipment has a comparatively limited range of uses. Unskilled workers can usually run the equipment without undue problems. However, automated equipment requires much higher skill levels to take full advantage of its capabilities. There are often unknowns and limited experience to overcome. Successful investment in and implementation of CIM requires creativity in software use and in gathering and assimilating information in order to realize the enormous capability gains it offers.

Intangible Benefits

It is much more difficult to quantify the impact on profits that can be realized with automated equipment than it is to quantify the impact of traditional equipment (Band and Hendricks 1987). This is because automating the factory brings many indirect and intangible benefits that are difficult to estimate.

1. While it is relatively easy to calculate the results of reducing labor costs, scheduling and set-up time, and works-in-process inventory, it is very difficult to calculate the benefits derived from improvements in quality, reduction in delivery time, increased flexibility, and competitive advantage.
2. It is comparatively easy to calculate expenditures on hardware and software, but difficult to estimate future operating costs of automated equipment because of insufficient experience.

3. It is difficult to predict the cost of CIM information systems.
4. The economic life of equipment is uncertain because of rapid technological innovation.
5. The potential loss in sales from failing to automate should not be overlooked, but this again is extremely difficult to quantify.

Wide-Ranging Strategic Implications

Investments in traditional equipment usually have limited strategic implications. However, investments in automated equipment and CIM have wide-ranging strategic implications. This means that investments in automated equipment and CIM should involve more people throughout the organization than investments in traditional equipment. Consequently, communication among the various fields of operations, manufacturing, engineering, information systems, production management, and accounting is essential.

Size of Investment

Traditional equipment and facilities investments are generally large scale in order to take advantage of economies of scale. On the other hand, investment in CIM is an "endless project" (Polakoff 1990). Instead of scale, economies of *scope* must be attained in CIM investment. As a result, investment in automated equipment tends to be undertaken on a small scale in many Japanese companies. However, the size of the investment is likely to be large because of the high cost of cutting-edge technologies. Consequently, investment in automated equipment must be analyzed very carefully.

Forecasting the Profits and Costs Related to Automation

The factors that are important to consider when investing in equipment differ depending on the degree of automation. Howell and Soucy (1987) classified the degree of automation into three levels: Level 1 (traditional equipment), Level 2 (FMS), and Level 3 (CIM). The corresponding factors for each level are shown in Table 8-1. As the level of automation increases, the intangible benefits become more important.

Table 8-1. Benefits Gained from Automation

Description	Level 1 (TP)	Level 2 (FMS)	Level 3 (CIM)
Object	Plant & equipment	Plant & equipment Software	Plant & equipment Software Network
Direct Profit	Labor reduction Energy conservation	Labor reduction Energy conservation Scrap decrease Inventory reduction	Labor reduction Energy conservation Scrap decrease Inventory reduction
Indirect Profit	Danger reduction	Danger reduction Support increase Space reduction	Danger reduction Support decrease Space reduction Lead time
Benefit		Quality Flexibility	Quality Flexibility Service to customer Competitiveness Throughput Learning effect

(Note: TP stands for traditional plant & equipment)

As indicated in Table 8-1, Level 2 equipment such as NC (numerical control) machines, unmanned transport systems, and unmanned warehouses requires investments in software. Level 3 (CIM) investments require building and maintaining information systems for networks and databases.

The direct benefits of Levels 2 and 3 are comparable. However, compared to traditional equipment, there are significant additional benefits in both quality improvement and inventory reduction.

Reductions in defects are realized due to the major role played by robots in improving quality. Further large reductions in inventories can be sought in Level 3.

Under FMS, support functions such as design, maintenance, monitoring, production planning, R&D, and software development must be beefed up. However, moving to Level 3 will decrease support staff requirements as design, maintenance, and other factors become automated. Lead time will be shortened through CIM integration of engineering and production systems with sales. Being able to respond to customers more rapidly improves the company's competitive position and is likely to result in more throughput. Considering the learning effect, conversion to CIM is an indispensable requirement for the next phase of automation, just as FA was the prerequisite for CIM.

Some costs and benefits remain easier to quantify than others. Table 8-2 lists some costs of CIM equipment and facilities. The costs that are easier to quantify include depreciation, interest, labor costs, costs for utilities (electricity, gas, etc.) and installation. The costs that are difficult to quantify include company training costs and system management costs. Generally, however, cost estimation is comparatively easy in contrast to calculating benefits.

Table 8-2. Cost Estimate for CIM

1. Quantifiable Costs
   - Cash flow of plant & equipment
   - Capital costs for plant & equipment
   - Labor cost
   - Utilities
   - Maintenance costs
   - Software costs
   - Others
2. Unquantifiable Costs
   - In-house training costs
   - Costs for system operation
   - Infrastructure
   - Others

The Benefits of Investing in CIM

There are significant advantages to investing in CIM. They include many indirect and intangible benefits in addition to direct profit. About 60 to 70 percent of the benefits of investing in CIM can be quantified in typical Japanese companies (Management Systems Technology Research Association 1989), but some benefits are difficult to quantify.

Reduction in Labor Costs and Energy Conservation

Labor costs are affected by CIM in two ways: reduction in shopfloor workers (direct labor) and increase in support functions (indirect labor). The clearest benefit of introducing FMS equipment is the reduction in shopfloor workers. The number of employees at Toshiba Tungalloy was reduced from 70 to 16, at Niigata Tekko from 31 to 4, and at Yamazaki Tekko from 204 to 10 as a result of introducing FMS (Japan Society for the Promotion of Machine Industry 1982). However, support costs such as maintenance, monitoring, product planning, design, R&D, and software development increase significantly.

As the degree of automation advances to FA from FMS, however, support costs also begin to drop. There is a reduction in the need for design personnel, for instance, because of the application of CAD. The introduction of CIM links engineering and production with marketing and aims for operations that can quickly adapt to environmental changes. Its focus is on reducing indirect labor costs and reducing cycle time. Equipment also becomes smaller in scale and more efficient as automation increases. As a result, energy is conserved. This should not be forgotten in the computation of the benefits gained through automation.

Reduction in Inventories

Work-in-process and finished goods inventories are significantly reduced as equipment is automated. This is accomplished by increasing flexibility in the production schedule, creating an orderly flow of goods, increasing quality, and improving schedule management. For example, in the previously mentioned case of Yamazaki Tekko, before automation, products were in process for three months, and the amount of money tied up was nearly ¥3 billion. After the introduction of automated facilities, work-in-process inventories were reduced to three days and the cost to ¥120 million. For these reasons, NEC includes reduction in inventories as a benefit when justifying automated facilities.

The reduction of inventory carrying costs and interest costs that result from reduction in inventories is another direct benefit gained from the introduction of CIM, as is the tax benefit from reduced inventories.

Quality Improvement

One outcome that can be expected at the initial stages of automation, along with a reduction in labor costs, is improvement in quality. This is because robots can perform simple tasks with a great deal of precision and reliability.

Quality improvement also has both easily quantified benefits and benefits that cannot be easily quantified. Reductions in the rate of defects and improvements in yields are easily quantified, while the increases in marketability that come from quality improvements are not.

Uniformity of production is possible with automated equipment and the number of defective products can be reduced. Improvements in quality can be quantified by decreases in the rate of defects and by increases in yields. For example, Toshiba Tungalloy increased its yield from 95 to 99 percent by automating its factories. It is easy to calculate how much profit was gained from this four percent increase in yield.

When products are made more uniformly, testing facilities and personnel decrease. The use of sensors grows and quality improves even further. When General Electric automated its dishwasher production facilities, customer service calls due to breakdowns decreased 50 percent (Kaplan 1986). As Kaplan points out, estimating the benefits from this type of quality improvement is comparatively easy.

These benefits are not the only advantages derived from improving quality. There are several others. Customer complaints decrease, the image of the company improves, and the products are easier to market. Though these intangible benefits are difficult to quantify, they should never be overlooked.

Reduction in Floor Space

Automated machines tend to be smaller in size and number than traditional equipment because of their greater flexibility and sophistication. Toshiba Tungalloy replaced 50 traditional machines with 6 automated machines. Niigata Steel decreased machines from 31 to 6 and Yamazaki Steel went from 68 to 18 machines. As a result, floor space set aside for machines can be reduced. The space needed to warehouse inventories of finished products and works-in-process will also decrease as the inventories decrease. The benefits of such decreases can be quantified.

Reductions in floor space are particularly important in Japan where real estate costs are very high. When calculating the effects of reducing floor space, this factor should be evaluated not only in terms of acquisition cost but in terms of current market value.

Shorter Lead Time and Throughput

CIM integrates production and engineering with marketing, and can therefore drastically cut lead time. For example, Yamazaki Steel shortened the mean processing time for one unit from 35 days to a day and a half. Another benefit of reducing the lead time is reduced inventories, which has already been considered above.

The intangible benefits from expeditiously satisfying customers are also important.

Increased Flexibility

Due to the introduction of automated equipment and facilities, products that could only be produced in large quantities by mass production can now be produced in high variety/low volume production with low cost. The competitive advantages gained from this flexibility are tremendous, although difficult to quantify.

It should not be forgotten that the flexibility of automated facilities provides an important backup capability. When a piece of equipment breaks down, products that were scheduled to use the machine can often be produced on another machine.

Because product specifications can be changed easily, products can conform with the needs of the customers. This is often practiced in automated automobile factories where the products coming down an assembly line differ from each other. In short, one of the advantages of using automated equipment and facilities is that simple changes in the flow can be made with ease.

Reduction of Dangerous Operations

The introduction of FA has decreased the number of dangerous and physically taxing operations performed by people. This has contributed significantly to the reduction of industrial accidents.

The first robots were designed to take over tasks that people did not like, such as welding. Following this trend, major Japanese general contracting companies are actively introducing robots to replace workers in the most dangerous work.

The reduction in the number of dangerous and physically demanding tasks is no doubt an important benefit, but again, difficult to quantify. Benefits include reductions in the rate of breakdowns, fewer absences due to injuries, and less special pay for hazardous work.

Improved Image for Manufacturing

Manufacturing has been haunted by the image of workers covered with oil and sweat performing simple tasks over and over again in an atmosphere of vibration and deafening noise. Young people, therefore, dislike the image of manufacturing which they label as the "three K's," *kitanai, kibishii*, and *kiken* (dirty, hard, and dangerous).

As a result, recruiting and retaining a manufacturing work force has become a major problem in Japan. For example, the best graduates from Tokyo University tend to work for banks or insurance companies rather than for manufacturing companies that have a dirty image. CIM can eliminate the less desirable aspects of manufacturing, and can provide a more agreeable atmosphere for work. While it is very difficult to quantify the benefit of this factor, it is very important in Japan.

The Learning Effect and the Competitive Advantages

Investments in new technologies have an important learning effect. Even if the direct cash flows associated with a particular investment are negative, investment may sometimes be indispensable just to maintain knowledge, and therefore, future competitiveness.

Considering the rapid pace of technological innovation, there is no future for companies that put off capital investments until new markets develop. Moreover, insofar as a company has not mastered a given level of automation, it cannot go on to the next level and will be left behind by the competition. Companies that did not have any experience with FMS in the 1970s were not able to install FA successfully at the beginning of the 1980s.

Table 8-3 compares the profitability of two investments: one in traditional equipment and another in CIM equipment. Three kinds of advantages are shown in the table: direct profit, indirect profit, and benefits. A direct profit of ¥566,000 can be gained by through introducing CIM equipment. There is an indirect loss of ¥17,000 because support labor costs such as planning and design, maintenance, and software development increase through introducing CIM equipment. Benefits could be quantified but many Japanese managers do not quantify them. They believe that CIM investment should be justified strategically by top management, taking these kinds of benefits into consideration. In sum, in this example, ¥549,000 and + β can be gained by introducing CIM investment.

In terms of the overall company strategy, investment in traditional equipment is suitable for pursuing the strategies of low cost leadership and/or differentiation identified by Porter (1985). Investment in CIM increases the company's ability to compete on a multidimensional rather than a uni-dimensional basis. Companies are simultaneously in a position to offer high quality product at low cost, and provide fast and flexible product changes. The ability to be able to compete on multiple bases has been identified as a key strategic imperative by a number of authors. (See for example, Hall 1980; Gilbert and Strebel 1988; and Loomis 1989.)

Choosing a Method of Investment Justification

After the benefits of installing automated equipment have been measured, one must compare these with the cost of investment and justify the investment.

The DCF Method and American Companies

In the United States, the most popular methods are based on DCF (discounted cash flow). Klammer (1972), Fremgen (1973), Gitman and Forrester (1977), and Kim and Farragher (1981) suggest that DCF methods, (in particular, IRR, the internal rate of return) are the most popular methods of evaluating investments in the United States. For example, in the 1981 study by Kim and Farragher, 49 percent of the companies used the IRR method while 19 percent used the NPV (net present value) method, 12 percent used the payback method, and 8 percent used the ROI (return on investment) method. DCF is now ranked first among

Table 8-3. Justification Table for CIM

(1 unit = ¥ 1 million)

Description	Traditional Equipment		CIM Equipment		Difference
Direct Profit					
Direct labor reduction	70 @ 4	280	5 @ 4	20	260
Energy conservation		70		30	40
Scrap decrease	90% @ 40	280	97% @ 40	80	200
Inventory reduction					
Work-in-process		60		15	45
Products		30		9	21
Subtotal		720		154	566
Indirect Profit					
Danger reduction		5		1	4
Support labor					
Planning & design	2 @ 10	20	3 @ 10	30	–10
Maintenance	2 @ 7	14	3 @ 7	21	– 7
Software development	1 @ 9	9	7 @ 9	63	–54
Space		40		12	28
Lead time		30		8	22
Subtotal		118		135	–17
Benefits					
Quality					
Flexibility					
Service to customer					
Competitiveness					
Throughput					
Learning effect					
Subtotal		+ α		$\alpha + \beta$	β
Grand Total	$838 + \alpha$		$289 + (\alpha + \beta)$		$549 + \beta$

Notes: 1. Danger reduction can be quantified by the spending for insurance.
2. Space can be quantified by rent.
3. In the case of traditional plant and equipment, the cost is ¥ 838 million and α benefit can be gained, whereas through CIM investment cost can be reduced to ¥ 289 million and $\alpha + \beta$ benefit will be gained.

capital budgeting elements by accounting experts (Church and Lambert 1993).

The DCF method is the most popular method among American accountants for automated equipment as well. For example, joint research carried on by the CAM-I and NAA in 1987 revealed that the IRR method was used by 69 percent and the NPV method by 64 percent (multiple responses) of companies evaluating investments in automated equipment (Howell, et al. 1987). The payback method was used more frequently when evaluating investments in automated equipment than when evaluating investments in traditional investment. However, the payback method is frequently criticized because "it does not consider all the project's cash flows or the time value of money, so its use could cause the company to make bad investment decisions" (Band and Hendricks 1987). It is generally agreed that the IRR and the NPV methods are superior to the payback and ROI methods for evaluating investments in CIM equipment (Polakoff 1990).

Overall Evaluation in Japanese Companies

Japanese companies do not normally justify investment solely on the basis of economic factors; they take other quantitative and non-quantitative factors into account. In the second Sakurai survey (1991) on FA and CIM conducted by mail in 1988, of the 198 companies that responded, 22 percent replied that they base their decisions primarily on considerations of economic factors, while 71 percent said that they consider other factors as well.

Since there are many factors in investing in CIM which cannot be quantified, in the final analysis, an overall evaluation that goes beyond quantifiable economic factors may be superior. For example, should investment in CIM be rejected in business practice only because the IRR does not exceed the cut-off rate? That would probably be best if the investment were in traditional plant and equipment. However, with CIM such intangible benefits as flexibility, improvement in quality, competitive advantage, and the learning effect can be of overriding importance. Thus, it may be preferable to use a process of decision-making that "puts practical business sense, not banker's logic, into the decision process" (Koenig 1990). An overall evaluation that incorporates strategic considerations is likely to be superior to an analysis that is based solely on quantifiable economic factors.

When Japanese companies do use a quantifiable method, they most often use the payback method. According to the Tsumagari (1972) and the Shibata survey (1988), more than 60 percent of Japanese companies use the payback method. In the Shibata survey, the IRR and the NPV methods were used by only 13 and 12 percent of the companies respectively. Note that in the 17 year interval between the Tsumagari and the Shibata survey, use of the payback method increased from 61 percent to 71 percent. Incidentally, both studies surveyed the same companies. In a January 1988 Sakurai survey (1991) on FA and CIM, nearly 74 percent of the companies replied that they used the payback method for justifying FA equipment.

Of course there are many drawbacks to the payback method. Its chief fault is that cash flows after the payback period are not taken into consideration, nor are the amounts or the timing of cash flows during the payback period. With faults like these, why do Japanese managers continue to use the payback method? Perhaps because it is easy to use. Since Japanese managers tend to like simple tools, this inference may be part of the answer. But this does not sufficiently explain why use of the payback method has continued to increase over time.

Mr. Koike, the statutory auditor of NEC commented that a two year payback period is appropriate for his company. According to him, capital that is invested in high technology products must be recovered as quickly as possible. It seems clear from his statement that at least at NEC, the payback method is used not because it is simple but because the goal is a quick recovery of capital. This important fact is not particular to NEC.

Thus, it is clear that many Japanese companies typically use the payback method. This raises an interesting question. Japanese managers are believed to have a long-term perspective (Kagono et al. 1985), yet they prefer the payback method which is based on short-term considerations. This differs completely from the approach of Japanese management as it is generally understood. The following reasons are offered for this apparent contradiction.

1. The life cycle of products has been shortening. The average life of a product in Japan is now two to three years (Nakane 1985). Thus, capital investment must be recovered quickly.
2. The life cycle of equipment has also been getting shorter. Sakurai's January 1988 survey shows that the life cycle of FA equipment in Japanese companies is declining. Out of 192 companies surveyed, the following were the capital recovery periods reported by the companies using the payback method: 2 years (6 percent), 3 years (35 percent), 4 years (10 percent), 5 years (33 percent), 6-20 years (14 percent) and others (2 percent). In short, 41 percent are recovering capital in three years or less and 84 percent are doing so in 5 years or less. These results are all bolstered by short term capital recovery.
3. Sales of high technology products cannot be accurately predicted. Semiconductors are a typical product of this type.

An Illustration of Investment Justification

How is investment justification in CIM actually made in major Japanese companies? In reality, it is difficult to select typical examples because procedures are different depending on the type of business. The following example of the investment in CIM by hypothetical company X is a composite combining the actual experience of several Japanese companies.

The Process of Decision Making in Investment Justification

Decisions for investing in CIM equipment involve four stages. First, proposals are generated by relevant departments in the mid-range business planning stage. Next, the framework for investment in plant and equipment is formulated in the policy making process for budgeting. In the third stage, coordination between the executive board and production planning groups are made through the budgeting process. Lastly, in the fourth stage final decisions are made about the investments.

Mid-range business plans extend over a two or three year period and are rolled forward each year. Within this plan, policies for expansion of operations, proposals for improvement, and other important issues are examined every year. In the first stage, a basic policy is determined, and studies concerning specific numbers are given secondary consideration. What is essential is to determine what action is to be taken. Policy making is more important than concrete numbers.

In the second stage, an investment proposal is initiated and studied on a companywide basis. A policy investment such as investment for environmental protection is considered separately from investments for expansion or improvement. Investment in CIM is considered to be largely a policy investment. In company X, it must be self-financed (financed only by depreciation and retained earnings).

In the third stage, a budget proposal is presented in the framework of profit planning. The need for and adequacy of the proposal in CIM equipment is reviewed in relation to the overall investment plan in the production department. This proposal is reviewed in conjunction with other proposals submitted by other departments. Here a method of economic justification such as payback, ROI, or IRR is applied.

In the final stage, after the budget has been determined, a *ringisyo* (formal proposal for top management approval) is submitted and final approval of top management is obtained. The research report on the investment proposal, and the proposal for investment criteria receive approval along with the budget proposal. The persons who make the decisions on the investments differ depending on the amount of money involved. For company X, investments must be approved by the following persons:

- More than ¥1 billion company president

• From ¥100 million to ¥1 billion	executive in charge (chief director, managing director, and so on)
• Less than ¥100 million	department head

The research report on the investment proposal includes the specifications, an outline (product, quantity to be produced, process, operating time, period of use), the reason for purchase (expansion, conservation of energy, rationalization of operations, reduction in manpower, quality improvement, reduction in environmental pollution, improvement in operations, renovation, and others), the conditions for purchase (from another vendor, import, specially ordered, valuation of similar item, and others), and the conditions for purchasing a substitute item. It also indicates why a specific vendor was selected, the results, the cost-effectiveness, and the future prospects.

CIM Investment Criteria

Decisions for investing in CIM equipment are not made independently of other investment decisions. Evaluation of benefits and costs from a investment are made in the third stage. The results are included in the document on equipment investment criteria, along with calculations of the cost of the new equipment, profitability, capital recovery period required, and comparisons with existing equipment.

Company X uses a two stage process to screen projects. In the first stage, investments are evaluated based on their projected payback and profitability. In the second stage, intangible benefits are considered. The primary criteria are shown in Table 8-4. The number in the lower line in each block is the weight given to that item.

The secondary criteria are shown in Table 8-5. These criteria include effects of the investment on quality, flexibility, and competitiveness.

Table 8-4. Primary Criteria

Description	Scores					
Profitability	20–30% 30	15–20% 25	10–15% 20	5–10% 10	0–5% 5	Minus 0
Payback (years)	1–2 30		2–4 20		4–6 5	Over 7 0

(Note: Numbers in the bottom line within each box are weights for the scores)

Table 8-5. Secondary Criteria

Description	Scores				
Future Outlook	Development 30	Growth 25	Mature 15	Saturation 10	Downward 5
Quality		Greatly 25	High 15	Moderately 10	
Urgency		Essential 25		Important 10	Convenient 5
Flexibility		High 20	Middle 15	Low 10	A little 5
Delivery			Very good 15	Good 10	Fair 5
Competitiveness			Very high 15	Strong 10	Fair 5

In the overall determination, investment proposals that exceed a score of 30 in the primary criteria and a score of 75 in the secondary criteria tend to be selected. Nonetheless, these are not automatically selected, and ultimately the person making the decision (the company president, the executive in charge, or the department head) will decide on the basis of corporate strategy.

Company X has factories in areas A, B, and C. Area A produces semiconductors, so strategic factors play a considerable role in investment decisions. Since technological innovation occurs rapidly and product life cycle is very short, long-term predictions are difficult. Therefore, in the primary criteria much greater weight is given to payback because early recovery of investment is indispensable for this industry. Because of the size and strategic nature of this investment, however, the intangible benefits and the judgement of the company president are likely to prove critical.

On the other hand, area B makes products in which there is hardly any room for technological innovation. Therefore, much greater weight is given to profitability analysis and the intangible benefits are far fewer.

In area C, the life cycle of products is short. Consequently, the payback period must be within two years and investments with a payback in excess of two years will be turned down even though the ROI may not be inferior.

For most projects at Company X after an investment has been made, the actual amount of investment, the amount of sales (the sales price times the volume), the variable production cost, the fixed costs of production, the variable marketing cost, the general administrative costs, and the effect of the investment on other business are carefully audited. Such post-investment audits are conducted each year by the internal auditing department.

Conclusion

Three new points become important with respect to investment justification of CIM. They are: 1) though it may be called investment in equipment, investments in CIM also involve investment in R&D and computer software; 2) many of the benefits of CIM are indirect and intangible and these benefits are very important in CIM investment justification; and, 3) there are differences in the methods of investment justification in the United States and Japan. Of these three points, the question of computer software will be covered in Chapters 9 and 10. In this chapter only the latter two issues were discussed.

With traditional equipment, it is often sufficient to evaluate investment only by quantifying the profits achieved by reducing labor costs and converting the profits each year to present values. However, when evaluating investment in CIM, one must take into account not only profits realized from reductions in labor-hours, but also indirect and intangible benefits realized from such factors as improvements in quality, reduction in inventories, shortening of lead time, flexibility, competitiveness, and so on. However, it is very difficult for business managers to include such intangible benefits in calculations. Ultimately, it is important for top management to make a final decision based on corporate strategy.

References and Further Readings

Band, Robert E., and James A. Hendricks. 1987. "Justifying the Acquisition of Automated Equipment." *Management Accounting* (July) p. 45.

Church, Pamela H., and Kenneth R. Lambert. 1993. "Outside Influences on Capital Budgeting Systems." *Journal of Cost Management* (fall) pp. 55-57.

Fremgen, James M. 1973. "Capital Budgeting Practices: A Survey." *Management Accounting* (May) p. 20.

Gilbert, Xavier, and Paul Strebel. 1988. "Developing Competitive Advantage." In *The Strategy Process*. Engelwood Cliffs: Prentice-Hall, pp. 82-94.

Gitman, Lawrence J., and John R. Forrester, Jr. 1977. "A Survey of Capital Budgeting Techniques Used by Major U.S. Firms." *Financial Management* (fall) p. 68.

Hall, William K. 1980. "Survival Strategies in a Hostile Environment." *Harvard Business Review* Vol. 58 (September-October) pp. 75-85.

Howell, Robert A., and Stephen R. Soucy. 1987. "Capital Investment in the New Manufacturing Environment." *Management Accounting*, Joint Study by NAA & CAM-I, NAA, pp. 145, 22, and 36.

Japan Society for the Promotion of Machine Industry. 1982. *An International Comparison of Management in High Technology Industries*. Japan Society for the Promotion of Machine Industry (56-3, March) p. 92.

Kaplan, Robert S. 1986. "Must CIM Be Justified by Faith Alone?" *Harvard Business Review* (March-April) pp. 87-93.

Kagono, T.I., K. Sakakibara, I. Nonaka, and A. Okumura. 1985. *Strategic vs. Evolutionary Management, A U.S.—Japan Comparison of Strategy and Organization*. North-Holland, pp. 37, 153.

Kim, Suk H., and Edward J. Farragher. 1981. "Current Capital Budgeting Practices." *Management Accounting* (June) p. 28.

Klammer, Thomas. 1972. "Empirical Evidence of the Adoption of Sophisticated Capital Budgeting Techniques." *Journal of Business* (July) p. 393.

Koenig, Daniel T. 1990. *Computer Integrated Manufacturing, Theory and Practice*. Hemisphere Publishing Corporation, p. 187.

Loomis, Carol J. 1989. "Stars of the Service 500," *Fortune* Vol. 119 (June 5) pp. 83-86.

Management System Technology Research Association, eds. 1989. *CIM: For Improvements in Management*. Nikkan Kogyo Newspaper Company, p. 130.

Nakane, Jinichiro (chairman). 1985. "Committee Report Concerning Medium and Long Term Prospects of Production Technology." *The Advance of Factory Automation and Future Concerns: The Medium and Long Range Outlook for Production Technology* (May) p. 13.

Polakoff, Joel C. 1990. "Computer Integrated Manufacturing: A New Look at Cost Justifications," *Journal of Accountancy* Vol. 169 (March) p. 28.

Porter, Michael E. 1985. *Competitive Advantage: Creating and Sustaining Superior Performance*. New York: The Free Press, pp. 12-16.

Sakurai, Michiharu. 1991. *The Change of Business Environment and Management Accounting*. Doubunkan, pp. 285, 326.

Shibata, Norio, and Kumata Yasuhisa. 1988. "Japan's Budget Control Systems: Survey of Present State and Future Issues." *Accounting* Vol. 40, no. 4 (April) p. 86.

Tsumagari, Mayumi, and Matsumoto Joji, eds. 1972. *Japanese Companies' Budgetary Control: Survey of Present State and Future Issues*. Japan Productivity Headquarters, p. 93.

CHAPTER 9

Cost Management for Software

Japan's manufacturing industries have achieved a level of development that enables them to produce quality products at low cost. One reason for this superiority is that they use approximately 60 percent of the robots in the world (Japan Robot Association 1994). They have also succeeded in reaching a high level of overall factory automation. But Japan's high level of manufacturing productivity is not solely the result of robots and other hardware. Without superior software, robots and other machines cannot be used effectively.

Thus, the role of software in Japanese companies increases as automation increases. For example, roughly 40 percent of the equipment cost for automating a factory is software cost. At one Mitsubishi factory where FA facilities are produced, software cost makes up 70 percent of the total cost of the FA facilities. The use of databases and networks characteristic of CIM further increases the relative portion of software cost. As a result, managing software cost is extremely important with FA and CIM production methods. Thus, the focus of cost management is moving from hardware to software.

Japan's Software Development Environment

Developing FA or CIM software is a complex and interdependent process. In addition to determining the needs of the customer, the needs of departments such as manufacturing engineering, product development, manufacturing, and marketing must also be considered. Deliveries must be made on time, and quality and service guaranteed. In this sense, the development of FA or CIM software is as Mr. Koike (an executive of NEC) described in 1991, "a joint activity of engineering, manufacturing, and marketing."

The Japanese software development environment differs from that in the United States in three significant ways. First, the Japanese place a high value on cooperative work. Second, they prefer custom software. Third, they often use software houses for software development. These characteristics come from the manner in which Japanese companies are managed and the special characteristics of their organizational structures.

Cooperative versus Individual Work

In the United States, originality is emphasized in software development. So, in contrast to Japan, American managers rely more on input by individuals than they do on the contribution of a team. American software development makes maximum use of individual creativity. As a result, requirements analysis and specifications analysis are typically done by individuals in the United States (Noguchi 1990).

Software development in Japan, on the other hand, is the result of the cooperative work of many software engineers. There is cooperation not only among members of the team, but from members of different departments as well. Since it is important for software developers to fully understand the needs of customers, they very frequently cooperate with each other. Thus, Japanese software developers analyze requirements and specifications as groups, not as individuals. In other words, one of the main characteristics in Japan is that they rely on the "creativity of the group" rather than on the "originality of an individual."

Custom Software

Japanese companies prefer custom software developed for a specific purpose rather than general purpose packaged software from external sources. For example, in the United States packaged software is more than 50-60 percent of all sales, while the percentage of packaged software in Japan was only 8-9 percent of all mainframe software development costs in 1994 (MITI 1994). Cusumano (1991) also commented that the major difference between the United States and Japan is that custom software and system integration accounted for the vast majority of Japanese software development. He noted that these two types of software accounted for 78 percent of the Japanese software market, compared to 39 percent in the United States.

The cost of custom software is high. Therefore, the Ministry of International Trade and Industry (MITI) has been trying to increase the use of packaged software in order to increase software reuse. However, packaged software use has not increased as much as hoped by MITI. While the cost of custom software is high in the short run, in the long run there can be higher productivity when using customized software since it meets actual needs. There is also the strong possibility that custom software will lead to better performance in FA and CIM where the level of complexity is so high that minor differences between the needs of the plant and the capabilities of the software can have large effects. In the United States, in contrast, selection of the appropriate packaged software and management of any necessary modifications dominates the typical software acquisition process.

Japan has little international market presence in the software field. Industry analysts remain divided over where Japanese firms currently stand in the software field (Cusumano 1991). For example, Imai and Ishino (1993) and Hayashi (1993) claim that the predominance of custom software has decreased efficiency in today's software development. Their discussion is very persuasive given the very bad business climate in the software industry. Furthermore, it is imperative for Japanese software houses to develop creative packaged software so that Japanese companies can pursue effective management, as discussed in Chapter 1. On the other hand, there are increasing reports of high productivity and quality in Japanese software. Additionally, there is the argument that the enormous creativity Japanese firms display in other areas of engineering should be of benefit to them in computer programming.

Both views are partly true. Japanese companies are strong in the *process* of FA applications software such as the steel industry, numerical control machine tools, and industrial robots. This is in contrast to the United States which is very strong in the *product* of packaged software as well as basic system software (for example, IBM and Microsoft).

Software Houses

Although software for FA and CIM can be developed by either user companies or software houses, in Japan much of this type of software is developed by software houses. These software houses tend to be subsidiaries of either the computer mainframe manufacturers, the software user companies, or independent subcontractors. Software is sometimes developed by employees of the user company itself. For example, Fujitsu has 52 subsidiaries and 116 independent software houses cooperating with it. NEC has 21 subsidiaries and 150 to 200 cooperating companies. Hitachi has 24 subsidiaries, and related companies with 18,000 employees (Kiryu 1986). Even though the users and mainframe manufacturers develop their own software, the software developer would normally be assisted by a number of software engineers from outside the company. For example, Tokyo Electric Power Co. employs 209 employees for software development but supplements them with more than twice that number from software houses.

This use of outside software developers would be very expensive in the United States, but in Japan it is just the opposite. In strong contrast to their American counterparts, Japanese employees do not like specializing—in anything—because it hinders chances for promotion. Even in management ranks the *tanoukou* (cross-trained) worker is the ideal. Those employees who do specialize migrate to specialist firms and accept lower pay. Thus, it is possible to use outside consultants at lower expense than internal employees. This method of securing services also solves the difficult personnel problems that would come from permanently assigning employees to specialist operations like computers, which would effectively end their careers.

Due to these three special characteristics of Japanese software development, cost management for software in Japan is focused on managing each software order as a whole (in contrast to departmental cost management), and these orders are often entrusted to external personnel.

The Characteristics of Cost Accounting for Software

Is cost accounting necessary for software and if so what kinds of cost accounting systems are necessary? We must know something about the characteristics of cost accounting for software to answer these questions.

The Software Industry is a Manufacturing Industry

Is the information service industry a service industry or a manufacturing industry? It is defined as a service industry by the

Ministry of International Trade and Industry (MITI), but whether it is really a service industry in the true sense is questionable. Many experts in Japanese software development believe the information service industry to be essentially a manufacturing industry. This is the major reason why the Japanese adopted the word "factory" for the place software is developed. Japanese companies have also adopted the factory approach to software management, consequently it is only natural that most software developers believe that a cost accounting system for software is essentially the same as that for industrial products.

According to a survey on cost accounting for software, more than two-thirds (77 percent) of the companies that develop custom software for others use cost accounting systems (Sakurai 1990). When software is to be used internally though, no more than one-third of the companies systematically calculate the development costs. Approximately half of the companies install cost accounting systems for packaged software products that they sell to others.

Software houses have most of the cost accounting systems. For products they intend to sell (packaged software), more than half the software houses and mainframe manufacturers and 40 percent of the user companies use cost accounting systems (Sakurai et al. 1993). In this survey, the questionnaires were sent to 1,169 listed companies and the 472 software houses in JISA. The response rate was 41 percent (675 responses).

The number of companies with some type of project costing system is lower in the United States than in Japan. Only one third (36 percent) of the companies surveyed in the United States (39 companies responded out of 50 independent software producers) have established costing systems (Coopers and Lybrand 1986). There seems to be no interest in cost accounting for software among American academics, as there have been few papers in the literature. In practice, only 33 percent of U.S. software houses have cost accounting systems. The oldest was installed in 1975, but most were installed in the mid- to late 1980s (Scarbrough, McGee, and Sakurai 1993).

What is the Cost Objective?

When building a cost accounting system, the first decision to be made is: What is the *Leistung* (cost objective) in cost accounting for software? Software is a new type of human creation that is closely linked with information. Historically, the chief area of research on cost objectives focused on physical products (since the 1875s when cost accounting was established), activities (since the 1950s), projects (since the 1960's)—and now, information (since the 1970s) (Sakurai 1981, 1987). Thus, we may need a completely different cost accounting system for this new type of economic resource.

Depending on the type of software and information service provided, software may fit into one of three categories:

1. tangible (for example, general purpose packaged software)
2. intangible (for example, maintenance or customer support)
3. combination of tangible and intangible characteristics (for example, custom software).

If the software is a pure intangible commodity, then it probably would require a totally different cost accounting system from what we have had in the past. However, software is best handled as a type of intellectual manufactured product. In this case, virtually the same cost accounting system can be applied as is used for hardware.

Characteristics of Software Costs

Software costs have some unique characteristics. One is that work-in-process has no physical substance. As a result, managing software costs is more difficult than managing hardware costs. In addition, it has become very important to properly measure the amount of labor-hours spent in developing software for cost management (Koike 1987).

Also, the cost structure for software is very different from that for hardware. Labor costs are very large, while material costs are very small. For example, the average labor cost in the 391 major Japanese software houses was 37 percent for in-house personnel (company employee), plus 23 percent for outside personnel (contract agreement). Since outside personnel (outsourcing or contract costs) are essential, total labor-related costs were 60 (37 + 23) percent of total manufacturing costs (JIPDEC 1990). On the other hand, the percentage of labor costs incurred by manufacturing companies in the first section of the Tokyo Stock Exchange in the same period averaged no more than 12 percent (Bank of Japan, 1990). In the report of manufacturing costs shown in Table 9-1, the labor costs for software development (employee and outside personnel) far exceed labor costs in typical manufacturing companies.

Table 9-1. Production Cost Report April 1, 1992 to March 31, 1993

Hitachi Information Systems **(Unit: ¥ 1,000)**

Cost Elements	Amount	Percent (%)
Material costs	1,605	3.3
Labor costs	12,257	25.4
Contract cost	23,599	48.8
Overhead	10,849	22.5
Total production cost	48,310	
Beginning WIP	1,930	
Total amounts	50,240	
Transfer to other A/C	6,263	
Ending WIP	3,393	
Total cost accounted for	40,584	100.0

Input-Output Relationship

In contrast to manufactured products, the input-output relationship is unclear in software development. The possibility that a software project will fail has not diminished significantly in recent years. In addition, there are big differences in the levels of skill possessed by software engineers. Thus, the output for any given investment is not certain.

Most Japanese software developers, however, believe that the input-output relationship is more stable than ten years ago because of the rapid technological development of software engineering. This is especially true in the area of custom software.

R&D-Type Projects

Some software development is similar to an R&D project. This is particularly true in system software or leading edge applications development. There is a high risk of failure in developing innovative software products.

Typically, however, most software development is not innovative. In most cases, developing custom software is like building a house. For these projects, there is a low risk of performance failure. Other projects are no different from industrial products. There is an extremely low risk of performance failure, and cost accounting can be applied easily and effectively. It is important to remember that risk in software development has been decreasing year by year in Japan.

Cost Accounting for Software

The first thing to consider when discussing cost accounting procedures is the selection of a method. Of the two main alternatives, job order costing or process costing, which is superior? Figure 9-1 illustrates the different forms of cost accounting for software (Sakurai 1987).

Choosing a Cost Accounting Method

Software can be reproduced in large quantities by copying the master. So, some argue, process costing is the appropriate

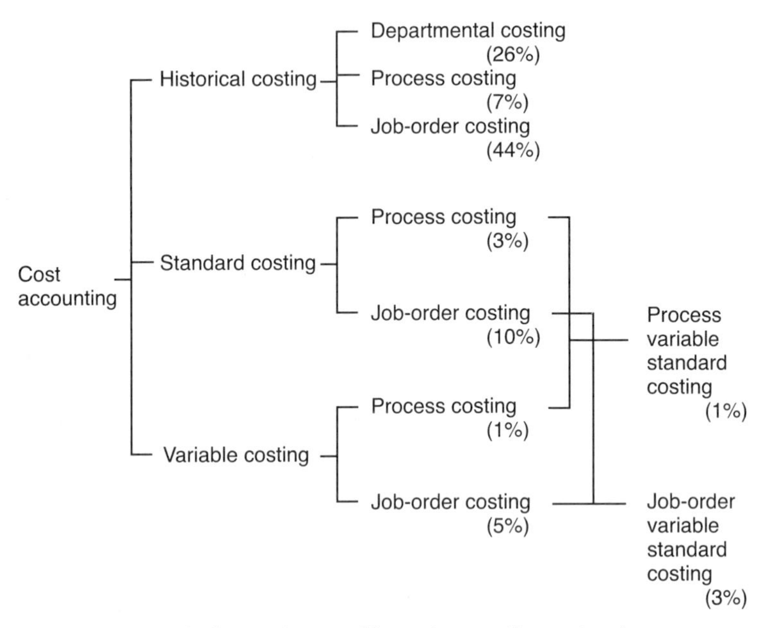

Figure 9-1. Cost Accounting Systems of a Software Factory

method. However, copying of the master is not when the value is added.

There is no need for us to apply a cost accounting technique to copying from the master because it is very easy to compute the unit cost of each copy. It is the production of the master itself that must be accounted for in mainframe computer software.

Each piece of software is fundamentally different from other pieces and places different demands on its creators. As a result, it is necessary to apply cost accounting techniques to each new piece of software. In process costing, inventory cost cannot be measured accurately because of its intangible nature. In other words, process costing requires some assessment of the percentage of completion of the program, which is difficult or impossible with software. Since job order costing has fewer requirements in this area, it is the preferred method for software.

Two-thirds of Japan's software companies use job order costing. In the 1987 Information Service Industry Association Survey, 62 percent of the companies (172 of 277) used it (Sakurai 1987). In the 1989 SOFTIC survey, 67 percent of the companies (113 out of 168) used it for software development (Sakurai 1993). In the 1994 Japan Personal Computer Association survey, 79 percent of the personal computer developers surveyed (39 out of 49) used job order costing (Sakurai 1994).

A significant number (one-fourth) of the software houses use still a third method: departmental profit and loss accounting. Though job order costing might be better, it is more expensive. The small software houses find it beneficial to use the simplest method possible. With the departmental profit and loss method, they can accumulate costs by project for each user.

Since departmental profit and loss accounting does not use a specific production order, it is not job order costing. However, since the software is not mass produced, it is not process costing either. It may best be referred to as "project cost accounting." This method is advantageous from a cost-benefit standpoint. It also has its disadvantages, among them, that work-in-process cannot be calculated automatically. However, when the evaluation of the work-in-process is not assigned a role in tax reporting or in financial accounting, this is not a serious problem.

Steps in Cost Accounting for Software

The steps in cost accounting for software are fundamentally the same as those used for industrial products. First, accumulation by cost elements (labor, costs paid to subcontractors, and overhead). Next, accumulation by departments, and third, cost assignment to products.

Cost accumulation by cost elements

According to the Cost Accounting Standards in Japan, the cost elements of a Manufacturing Costs Report should consist of materials, labor, direct expense, and overhead. These Standards were originally developed by the Ministry of Finance for manufacturing companies. The cost elements seen in the software industry are quite different from those seen in the manufacturing industries in two ways. First, there are virtually no direct materials used in developing software. The most typical cost element in direct expense is the costs paid to subcontractors. As a result, many software developers report on costs using the three cost classifications of labor, costs paid to subcontractors, and overhead. For example, CSK (one of the largest software houses in Japan) uses these three categories in their external financial statements. Many software developers exclude the cost of materials from reports on manufacturing costs.

Cost accumulation by departments

Cost accounting by departments was not widely practiced by Japanese software houses in the past because they were often small companies with less than 30 employees. Nevertheless, it has gained popularity, and by the late 1980s, 78 percent of the houses used it (Sakurai 1987). Cost accounting by departments should be the basis for software cost management because it provides cost control and accurate cost accounting. The fact that costing by departments has been adopted by many companies is a logical result of cost management practices needed in an intensely competitive environment. Moreover, the necessity for costing by departments will increase in the future.

The departments at a software developer differ significantly from those of hardware developers. For example, typical manufacturing departments for software developers are systems development, engineering development, basic design, detailed design, computer processing, operations, and so on. Some typical service departments are training, R&D, systems research, TQC, data management, TSS (time-sharing system) management, and so on.

Cost assignment to a product

The typical cost objective for industrial products is the product, and the cost unit is the job, unit, t, kg, lot, batch, order, or contract. On the other hand, the cost objective for software is normally a system or a program, and the cost unit is the project, the customer, the contract, or even the job.

When adopting job order costing, a job order must, of course, drive the system. The time ticket for labor-hours then becomes the basis of cost management. Direct costs are directly assigned and indirect costs are allocated to each project. The allocation method for overhead is fundamentally the same as with manufacturing cost accounting. More than two-thirds of the software houses use a predetermined application rate for allocation of overhead. The most commonly used allocation bases are direct labor-hours (36 percent) and direct labor costs (27 percent) (Sakurai 1987). While the software industry is still labor intensive, there is no discussion of other allocation bases, such as machine-hours. As the software industry turns into a capital intensive industry, many Japanese managers will switch to machine-hours. Recently, some of them have introduced ABC (activity-based costing) into the software industry.

One driving force behind the cost accounting systems is that many software developers need to provide full cost information for reimbursement under common cost-plus contracts.

Only 18 percent of developers use direct costing (variable costing), and the remaining 82 percent of the software houses use absorption costing. The form of absorption costing is more extreme than would be seen in American manufacturing systems. The Japanese calculate the full absorption cost by adding sales and general administrative expenses to manufacturing costs either regularly (51 percent) or when necessary under production contracts (31 percent). This practice differs markedly from manufacturing companies, which do not continually calculate full absorption costs. In the manufacturing sector one of the major reasons for cost accounting is to provide product cost information for preparing financial statements. Product costs calculated for this purpose would not include non-manufacturing costs such as sales and general administrative expenses.

Cost Management Tools for Software

The major tools for cost management in software development are standard costing, budgeting, target costing, and TQC.

Standard Cost Accounting

Software development is basically labor intensive work. This suggests that standard costing can be an effective tool for controlling development cost. In fact, many Japanese mainframe manufacturers use standard costing for controlling software costs. However, standard costing is not widespread for other software developers. Only 13 percent of the software houses make use of standard costing.

Time and motion studies which are effective in industrial production cannot be applied to software development because developing software is an intellectual operation that changes from project to project. Thus, Japan's mainframe manufacturers often use cost models to set standards instead of attempting to use work measurement methods. Japanese mainframe manufacturers, who have extensive experience in developing systems software, have a great deal of data to use in formulating cost models. On the other hand, typical user-developers and the smaller software houses do not yet have sufficient data to be able to create cost models.

The setting of cost standards and variance analysis for software differs from that of hardware. Specifically, there is no need to designate a standard cost for direct materials. The most important part of software standard costing is direct labor costs. For example, the following standard is set for direct labor costs, input costs, and machine costs for a typical software developer.

$$\text{Standard direct labor costs} = \text{the standard wage rate} \times \text{the standard labor-hours}$$

$$\text{Standard direct input costs} = \text{the standard entry unit cost} \times \text{the standard entry hours}$$

$$\text{Standard direct machine costs} = \text{standard CPU rate} \times \text{the standard machine hours}$$

Several factors hinder the adoption of standard costing. First, software is not a mass-produced product, but it is produced individually. Thus, setting a standard cost for each program takes time and is very difficult. Second, standardization is difficult because the skills of systems engineers vary widely across individuals. Third, software engineering techniques are improving rapidly. Additionally, once a standard is set it becomes behaviorally fixed and work group flexibility is likely to be lost. This could threaten the future development of software technology, and would certainly slow it down in many companies. Standard costing is thus a two-edged sword, which must be considered carefully before implementation.

Budgeting and Progress Management

The core of the manager's control of software development is self-regulated management with profit goals and cost targets. In other words, management sets the overall profitability targets, but leaves the details to the lower level managers and employees. It relies on the quality and training of the employees to innovate their way to good performance. Software developers also use profit planning and budgeting for this purpose. A profit plan should first be prepared, cost targets set next, then highly efficient software development activity conducted to achieve the profit goal and cost targets. Analysis of performance should be made between the results that could be obtained and the goal and targets.

To illustrate the relationship between profit planning and cost targets we will look at Melcom Service Company. Melcom develops its profit goals at the profit center, as shown in Figure 9-2, and cost targets are assigned to the cost centers. Performance is measured based on those centers.

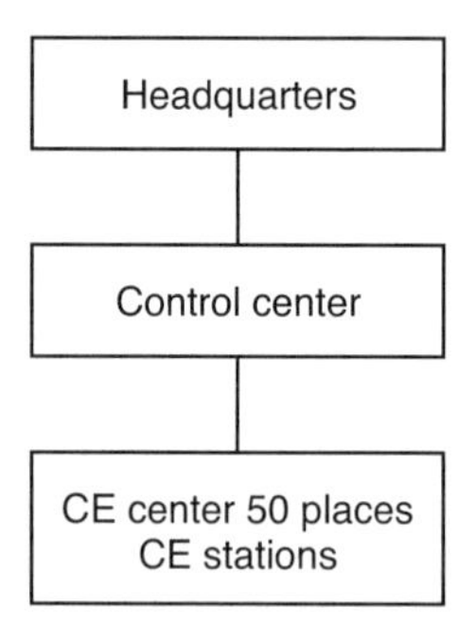

Figure 9-2. Organization and Profit/Cost Goals

The control centers of Melcom Service are located in Sendai, Tokyo, Nagoya, Osaka, and Fukuoka. Profit goals are assigned to these profit centers and operations are based on these goals. On the other hand, cost targets are assigned by the Customer Engineer center. Improved productivity, quality, operational control, and time management are sought in these profit centers.

The chief goal of this company is to complete the activity efficiently relative to the given profit. In a situation where individual orders are produced, like software development, periodic profit control with budgets is emphasized along with cost control for each job. An important issue in budgetary control for software is the comparison between periodic reports of actual costs and the budget. Each cost item is totaled in weekly or monthly units, recorded in tables, and a variance analysis with the budget performed.

It is necessary to check the progress of development work against the schedule by using a budgetary control trend chart.

The comparison between the budget and results should be conducted at the overall project level, the intermediate levels (subsystems, such as the Inventory Module in an accounting program), and for sub-sub-systems (such as the Inventory Module Help System). Figure 9-3, a cost item control table, is an example of a trend chart.

Month, Year / Cost	From April 199X to March 199X		
	April 199X	Cumulative Amount	Forecast based on Actual
Labor cost			
Purchase cost of software			
Contract cost			
Computer costs			
Supplies			
Travel expenses			
. .			
. .			
. .			
. .			
Total			

Figure 9-3. Control Chart of Cost Elements

Linking cost management to progress management is important to budgetary control for software development. Budgeting plays an important role in effective project management for most developers. However, progress management with the budget system is not as easy as it might seem.

When product development is behind schedule, even in only one module, the entire operation will probably be delayed. The delay of product development means more labor and other costs than were planned. In addition, changes in the external environment (mainly in the marketing environment, such as changed lead times, features, or operational requirements) and changes in internal environment (mainly management conditions such as transfers of personnel, a new project, delay in operations, or lack of appropriate experience by software engineers) will make changes in planned costs inevitable. In this situation it is very important to determine what the effect on profits will be by always comparing actual with planned figures. It is no exaggeration to say that progress management for each operation is the heart of cost management for software.

It is sometimes difficult to estimate sales amounts and costs when developing a part of software such as system design, detail design, and testing. In such cases, the estimated costs may change due to changes in design and operating requirements. Normally, software developers have hundreds of orders and for most projects personnel will be reassigned regularly due to changes in shipping dates desired by customers, delays in the development of hardware, or delays in internal company operations. Consequently, planned costs may change. Thus, it is necessary to periodically (usually every three months) compare actual costs with the budget, and then to modify and revise the budget. It is also essential to trace the variations in costs in the budget systematically from the time of estimation to the time of actual shipping, and to identify the cause of the variations. Typical reasons include increases in costs from mistaken estimates and failures in operations due to lack of personnel expertise.

Target Costing for Computer Software

Target costing is an effective tool in cost management for software. When a software contract is discussed with a customer, the developer forecasts the sales price. From the forecasted sales price, target profit can be computed. Japanese software development houses typically use return on sales (ROS) goals for deriving the target profit from a forecasted sales price. Allowable cost can be computed by subtracting the target profit from the forecast sales price.

Systems engineers determine the cost of development through cost studies and analysis, and set an estimated cost (called drifting cost). A cost model is sometimes used in the process of computing the estimated cost. The cost model is a way of predicting estimated cost using past experience as a guide. Such models take the following form:

$$C = f(v)$$

Here C is the cost for software and v is the parameter that drives cost. As shown in Figure 9-4, these parameters include programmer ability, product complexity, product size, available time, required reliability, and level of technology (Fairley 1985).

The first task in creating a cost model is to analyze and describe the characteristics of the person who will be doing the work. Specifically, clear understanding of the skill and experience (including academic record and qualifications) of the software developers, as well as the resources that the developer may use (time, supporting software, devices, etc.) are necessary. Next, develop an understanding of the desired performance of the project, the scale, complexity, limitations on development, and hardware limitations as characteristics of the project.

Then, derive the target cost from the cost model, and make the target cost the basis of the standard cost. The inputs to the cost model and its use can be depicted as in Figure 9-4. Since software has a tendency to be specialized, it is necessary to designate a cost model for each separate version (Koike 1987).

While calculating the estimated cost for each feature and process, the goal of cost reduction is always present. After cost reduction activities have been tried the drifting cost is calculated and compared with the target cost. When the drifting cost is larger than the target cost, it is once again subjected to cost reduction activity. If the drifting cost cannot be lowered to the target, then negotiations to raise the price may be necessary. In calculating the drifting cost, modules of code which can be converted or made into general purpose modules will be taken into account. The ability to convert or reuse parts of an existing program is a major point where cost management for software and hardware differ (Matoi 1987).

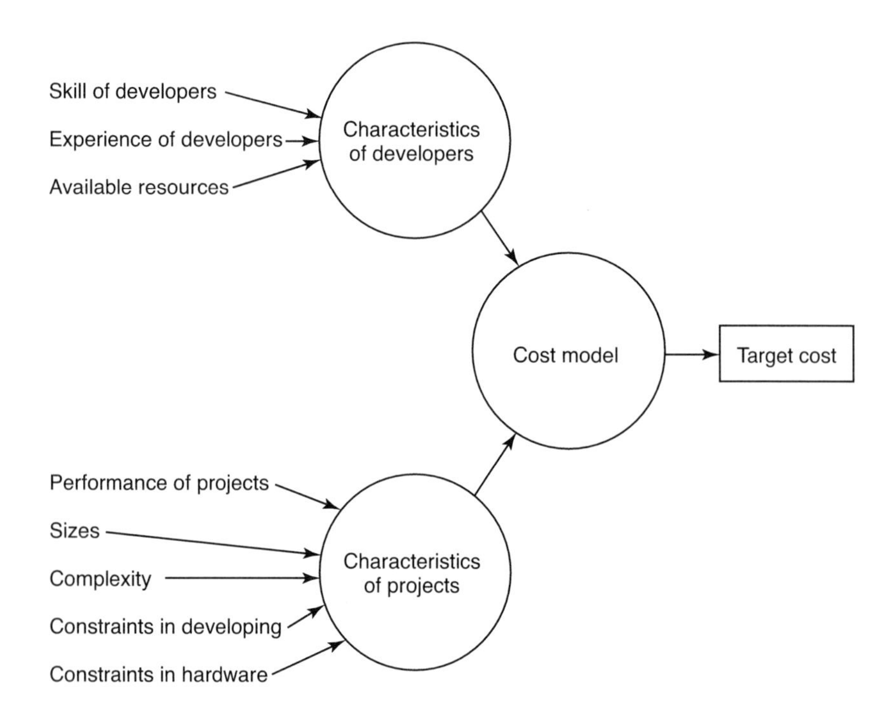

Figure 9-4. A Cost Model and Its Use

TQC (Total Quality Control)

The most important aspect of cost management in software development is improved quality. Since software in Japan is developed through teamwork, TQC is a very effective means of managing it. TQC was introduced to Japanese manufacturing companies at the beginning of the 1970s but its use in software development came much later, in the early 1980s. According to Mr. Kobayashi, the Chairman of NEC, the easiest thing to understand as an objective of quality in software development, and the easiest thing to do, is to eliminate bugs (Kobayashi 1982). TQC for software began around 1980 in Japanese mainframe manufacturers. From about this time both NEC and Toshiba aggressively developed unique types of TQC. In NEC it is called SWQC (Software Quality Control).

Japanese software houses began TQC much later than mainframe manufactures. The manner in which TQC proceeds depends on the company. The following is an example of TQC which focused on "composing sentences," or improving writing quality. It is from Nihon Denshi Kaihatsu, one of the subsidiaries of NEC.

Nihon Denshi Kaihatsu began TQC in 1983 with the following three aims: 1) to make the company more competitive, 2) to heighten value added by increasing labor efficiency, and 3) to develop an entrepreneurial spirit for developing better software (Sato and Sakurai 1986). What they began is referred to as SHOOT (Software House-Oriented Optimum Technique). SHOOT consists of increasing the efficiency of design, decreasing the number of bugs, rationalizing clerical operations, and increasing the technical level and training of personnel.

They realized that one of the products of software development was the written word. There were frequent complaints from customers that the language in the documentation was poor, and that it reduced the ability to use the product. In order to address this problem the company began an organized effort to improve documentation. In fact, correction of the problem took an extensive amount of time and was technically difficult. Managers assembled some examples of poor writing in company material, including documentation, memos, letters, contracts, and promotional material. They met with some 200 personnel above the level of manager once a week for two weeks to examine and make corrections. Then, they made three levels of evaluations—A (very good), B (good), and C (fair). This activity was continued for three years. This, of course, drew enormous attention to written communications in the company. The company has now developed and marketed a software package of computer aided instruction courseware called "How to Write Sentences." In addition, they have linked this with the office automation of individual labor-hours record tables, which they have also developed and sold as courseware.

Toshiba Engineering began a TQC called ETQC in 1983 developed from Zero Defects (ZD). They introduced ZD in 1978 but it was unsuccessful because the company is not a manufacturer, but a sort of engineering company. The company wanted to have a quality control tool suitable for the engineering industry. The original meaning of the "E" of ETQC was engineering, but it also refers to the amalgamation of top-down and bottom-up. It also suggests the developmental goals of organization. There are four slogans in ETQC: 1) to raise the skill level of each employee, 2) to make each workplace a creative environment, 3) to develop know-how as an engineering company, and 4) to perform each task easily, correctly, expeditiously, and safely at a low cost. The ultimate goal of ETQC is raising the intellectual level of employees by improving training. In fact, ETQC has been very useful in increasing the productivity of knowledge-intensive work (Toshiba Engineering 1987).

Conclusion

In this chapter we have compared the characteristics of Japan's software development with that of the United States and then given an overview of cost accounting for software in Japan. In order to effectively manage software costs, it is essential to establish a cost control system. Options discussed include standard costing, budgeting, target costing, and TQC.

Standard costing is important for cost control for software, and what is most important for Japanese companies is project management with the budget. In developing software, work delays lead to delays in shipping that ultimately raise costs. There is no management measure now available which is superior to budgetary control in identifying rising costs.

Another tool which is promising for cost management for software is target costing. At present there is no established implementation doctrine describing target costing for software development. Yet, when one considers the possibilities for management, target costing has the potential to become an indispensable tool for strategic cost management since it can focus the different abilities and experiences of the workers on the goal of profit planning and overall cost reduction—a goal subtly but critically different from traditional tools.

The human resource management elements that influence software development are also very great. Improving individual abilities with training and teamwork is indispensable in solving the problems of individual differences in ability. TQC, which most effectively promotes self-education and mutual education, is an activity most appropriate for Japanese companies, and absolutely essential in software development.

A final comment on cost management for software concerns the focus of Japanese managers on cost management issues, not on performance evaluation. The main aim of cost accounting for software is on cost management which is the greatest concern of Japanese managers, while measurement or resource allocation seems to be the major concern of American managers. We will discuss chargeback systems for measuring and managing software development and information processing costs in the next chapter.

References and Further Readings

Bank of Japan. 1990. *Analysis of Operations in Major Companies*. Finance Ministry Printing Office, p. 4.

Coopers and Lybrand. 1986. *Accounting for Software Development Costs: An Implementation Guide*. Coopers and Lybrand (U.S.A.) p. 24.

Cusumano, Michael. 1991. *Japan's Software Factories, A Challenge to U.S. Management*. Oxford University Press, pp. 51, 47, 6-7.

Fairley, Richard E. 1985. *Software Engineering Concepts*. McGraw-Hill, p. 65.

Hayashi, Ryozo. 1993. "Software Industry in Japan and Its Policy." *Business Review*, Hitotsubashi University (Institute of Business Research) Vol. 41, no. 1, pp. 21-22.

Imai, Kenichi, and Hukuya Ishino. 1993. "Software in Japan: Reasons of Backwardness and Remaining Problems to be Solved." *Business Review*, Hitotsubashi University (Institute of Business Research) Vol. 41, no. 1, p. 8

Japan Robot Association. 1994. *The Present State of and Outlook for Industrial Robots*. (October) p. 16.

JIPDEC (Japan Information Processing Development Center). 1990. *Report of Survey of the State of Management in the Information Processing Industry*. (March) p. 45.

Kiryu, Hiroshi. 1986. *The Real Picture of the Software Industry*. Nikkan Kogyo, pp. 77-86.

Kobayashi, Koji. 1982. *C & C and Development with Software Personnel*. Simul Publishing, pp. 148, 154.

Koike, Akira. 1987. *Cost Management in the Age of Software*. NEC (Nippon Electric Cultural Center), pp. 66, 115, 122.

MITI (Research and Statistics Department, Minister's Secretary). 1994. *Survey of Selected Service Industries, 1993—Information Service Industry*, MITI, pp. 24-25.

Matoi, Yasuo. 1987. "Software and Cost Control," *Journal of Business Practice* (June), p. 10.

Noguchi, Tasuku. 1990. *Business Administration of Software.* The Moriyama Bookstore, p. 170.

OECD. 1985. *Software: An Emerging Industry.* Organization for Economic Co-operation and Development (Paris), p. 83.

Sakurai, Michiharu. 1981. *Research on American Management Accounting Standards.* Hakuto Syobo Publishing Company, pp. 26-28.

Sakurai, Michiharu et al. 1987. *Cost Accounting for Software.* Hakuto Syobo Publishing Company, pp. 36, 191, 198-199, 201, 250. Sakuari chaired the committee on cost accounting for software.

Sakurai, Michiharu. 1993. "Survey on Accounting Practices for Software." *Accounting* Vol. 42, no. 7 (July), pp. 91, 93.

Sakurai, Michiharu et al. 1993. *Accounting for Software.* Chuokeizai-sha, pp. 332-333.

Sakurai, Michiharu. 1994. *Analysis of Mail Survey on Accounting for Software.* Proceeding of a Tentative Paper at Japan Personal Computer Association, August 8, 1994. Sakurai moderates all three committees (Information Service Industry Association, SOFTIC, and Japan Personal Computer Association) on accounting for software.

Sato, Koji, and Michiharu Sakurai. 1986. "Cost Management in Software Development Companies." *Journal of Business Practice* no. 385 (April/May), p. 19.

Scarbrough, Paul, Robert McGee, and Michiharu Sakurai. 1993. "Accounting for Software Costs in the United States and Japan: Lessons from Differing Standards and Practices." *The International Journal of Accounting* Vol. 28, p. 310.

Toshiba Engineering (Ltd.) Committee on Advancement of ETQC Activity. 1987. *Improving Intelligent Productivity Written by a Group of Engineers.* Kanki Publishing Company, p. 44.

CHAPTER 10

Chargeback Systems for Computer Resources Consumption

Software developers, like other business managers, must know the cost of their computer software in order to set prices effectively. Cost information is also essential for controlling cost. In response to these needs, Japan's major mainframe computer manufacturers and software houses have installed cost accounting and cost management systems for computer software. As was reported in chapter 9, three-quarters of Japan's leading software developing companies have introduced cost accounting systems for software (Sakurai 1987, 1990).

Some software user companies not only use and maintain software developed by the software houses, but also develop their own software, which they sometimes sell to other companies. These kinds of companies are increasing in number as software engineering expertise penetrates the economy. The special cost control problems of software development activity are now being addressed through the introduction of cost accounting systems.

How should user/developers control their costs for internal information processing? How should software developers whose chief mission is developing and selling software control their costs when they also provide information processing services to other departments within the company? One answer to managing the cost of the information processing department is the chargeback system.

The Chargeback System in America and Japan

In the United States, cost accounting for software is not common but the chargeback system is widely used. According to surveys conducted by Solomon and Tsay (1985) and McGee (1987), 75 to 85 percent of the companies used the chargeback system. When the author visited leading American companies for a survey of software cost control practices in 1989, he found that top-flight companies such as Westinghouse and Boeing had adopted the chargeback system, even though most of them had no cost accounting systems.

Cost Accounting for Software versus Chargeback System

In Japan many companies use cost accounting for the cost of developing software, while the chargeback system is used for the costs of information processing. A number of Japanese user companies employ the chargeback system for information processing inside the company, whereas cost accounting for software is mainly used by software houses that develop software to be sold to user companies.

In order to make the characteristics of a chargeback system clear, cost accounting for software and a chargeback system are contrasted in Table 10-1.

The chargeback system can be applied to the costs of software development as well. It is used not only by software users but also

Table 10-1. Cost Accounting for Software and the Chargeback System

Types of Systems	Cost Accounting for Software	Chargeback System
Main purpose	Software for sales	IS for internal users
Main object	Development cost for software	Expenses for IP
Main users	Software developers such as software houses	Software users such as auto makers

(Note: IS = Information system, IP = Information processing)

by software houses and manufacturers of mainframe computers. As Table 10-1 shows, it can be effectively applied for measuring the performance and controlling the cost of information processing departments. The main type of companies that use chargeback systems are software users. In certain companies, as shown in Table 10-2, cost accounting systems for software are applied to the costs of developing software, while the chargeback system is applied to the cost of information processing.

In Table 10-2, the terms *production*, *sales*, and *business administration* are abbreviations for software that is used for production management, sales, and administration respectively. Some Japanese companies treat all information processing costs together and do not have cost accounting systems for software. Other companies differentiate between the software for sales and the software used for in-house information processing. Most software houses only use cost accounting systems for software and do not have a chargeback system.

Chargeback System in Japan

There are correlations between cost accounting for software and chargeback systems in Japanese companies. According to the author's survey of user companies, not software houses, in January 1992, more than half (51 percent) of the companies surveyed

Table 10-2. List of Total Software Costs

Purpose of Development / Types of Development	Costs for Software Development							Costs for IS	Total
	R & D	S/W for Sales			Internal Use				
		Ps	0	B	Pr	S	B		
Inhouse development Labor cost Contract cost Other overhead									
Contract arrangement Software development Consultant									
Purchase									
Total									

(Note: Ps = packaged software, 0 = order made, B = embedded software
Pr = production, S = sales management, B = business administration)

(146) have neither cost accounting systems nor chargeback systems. Twenty percent of the companies have no cost accounting systems but have chargeback systems with budgetary control tied to the financial statements. On the other hand, of the companies with cost accounting systems, 60 percent have *both* cost accounting systems and charge back systems with budgetary control tied to the closing of accounts.

In short, the chargeback system is used by some manufacturers of mainframe computers (among them, Hitachi), computer user companies (such as Matsushita), and some software houses (such as Assist) in Japan. However, these companies are in the minority—not many companies use chargeback systems in Japan.

Instead, Japanese companies have used cost accounting to aggressively and independently systematize the control of costs in the information processing department. Two main reasons are given for not using a chargeback system: 1) cost accounting for software is widely used for cost management, and 2) unlike cost accounting for software, the goal of a chargeback system is performance evaluation. Few Japanese companies place as much emphasis on employee or unit performance evaluation as their American counterparts do.

To summarize the results of the surveys done in January 1992, we found that only 40 percent of the Japanese companies use the chargeback system. (Within this 40 percent, 25 percent used chargeback systems integrated with budgeting and closing accounts, 15 percent applied chargebacks only in the budgeting process, and 12 percent allocated information processing costs to user departments.) Of the remainder, 43 percent treated information processing costs as overhead without any allocation, while managers at 5 percent of the companies did not know what a chargeback system was (Sakurai 1990). Another survey supported these results (JIPDEC 1992).

Accounting for Information Processing Costs

The costs of an information processing department may be handled in one of two ways. They may be treated as overhead of the information processing department, or they may be allocated or transferred to other departments.

Information Processing Costs as Overhead

The information processing department is either considered part of the company's head office or is considered part of the factory. If treated as part of the head office, its costs are treated as general administration overhead. If treated as part of the factory, its costs are treated as factory overhead. These costs are not assigned or transferred to the operating departments for budgeting purposes.

Information Processing Costs Transferred

Costs for information processing are often allocated to a primary cost center or profit center in both the United States and Japan. According to a survey conducted in the United States by Fremgen and Liao in 1981, 89 percent (full allocation, 62 percent; partial allocation, 27 percent) of the companies allocate the costs of the data processing department for performance evaluation. The Japanese result was 75 percent (87 out of 116 companies) according to a survey by Nishizawa (1991). However, this does not mean that those companies that transfer or allocate information processing costs to other departments use a chargeback system.

When information processing costs are transferred to operating departments, there are two treatments for these costs. One is to allocate them to operating departments by some formula that is unrelated to the actual usage of information processing resources. The other is to treat the information processing department as a service center in its own right and transfer its costs to departments consuming its services on the basis of usage. This second method is defined as the chargeback system.

The Chargeback System

A chargeback system allocates costs to user departments. Here "allocate" expressly means transferring the costs to the user. When transferring costs, the charge is assumed to occur on a one-to-one basis between "buying" and "selling" sub-units (Choudhury et al. 1986). As McGee (1986) has rightly pointed out, it is similar to the manner in which a transfer pricing system works. In the chargeback system there is a need to design the system so that it is closely linked to policies on divisional autonomy, objective performance evaluation, and goal congruence between the head office and the divisions.

There are basically two methods for transferring the costs of an information processing department to the user departments. One is called the cost centered approach (also called the service department approach), and the other is called the profit centered approach.

The cost centered approach

The cost centered approach treats the information processing department as a cost center. It is based on the principle that a portion of or all of the costs incurred should be covered by the other departments that use its services.

The advantage to the cost centered approach is that the costs of the information processing department are effectively expended. That is to say, demands for unnecessary data processing that are generated merely because the service is free will not be performed if the cost will be assigned to the user department. In addition, placing a charge on the service can heighten the interest in information processing by departments that use the information processing service.

The disadvantages of a cost centered approach are that it increases clerical procedures and may discourage use by departments that have significant needs for the service but do not have ample resources.

The profit centered approach

The profit centered approach treats the information processing department as a profit center and transfers the amount of cost covered plus an added profit to the department that received the service. Users can be from inside or outside the company. In the latter case, cost accounting for software is more effective for pricing and billing.

Besides using the cost for the price that is transferred, the market price is transferred to the department using the service when possible. The goal of this approach is that, when cost effective, information processing should be accomplished outside the company. However, it is difficult to ascertain what the market price is, and in fact there are many cases when the market price cannot be determined (MAP 1987).

The profit centered approach reduces the costs of the information processing department more effectively than the cost centered approach. As a result, the profit centered approach is believed to be theoretically superior to the cost centered approach.

Chargeback Systems in the U.S. and Japan

There are several surveys on chargeback system in the United States. According to a study conducted by Choudhury et al. (1986), more than 80 percent of American firms replied that the profit centered approach is undesirable. Choudhury attributed the limited use of the profit centered approach to the fact that the information processing departments have not yet reached a high stage of development.

Similar results were obtained by McGee (1987). McGee's survey of American companies (sent to Fortune 500 companies in a mail survey to which 450 replied), the great majority (84 percent) of these companies use the chargeback system. Very few companies (16 percent) treated the costs of the information processing department as overhead. The greatest number of companies using the chargeback system treated the information processing department as a cost center. Only a handful of companies treated it as a profit center: 7 percent in the case of software development costs, and only 8 percent in the case of operating costs.

Another survey conducted by Solomon and Tsay (1985) (Fortune 500 companies, 185 replies), similar to the study done by Choudhury and McGee, shows that the great majority of the companies (78 percent) surveyed used the chargeback system. Only 21 percent treated the costs of the information processing department as overhead. Interestingly Solomon and Tsay found a greater likelihood of chargeback system use in more decentralized companies. In those companies, the responsibility for decision making rested at lower levels. Those companies having an information processing unit as a profit center were more likely to use a chargeback system. From the above, we can generalize that most American companies use the chargeback system and implement it in the context of cost centers.

A comparative study made in January 1992 revealed that only 40 percent of Japanese listed companies used the chargeback system. Nearly all of the companies that use this system rely on the cost centered approach. That is, only 10 percent of 63 companies that use the charge back system use it with profit centers. This low use of profit centers is similar to the United States. Of the 14 companies the author visited in 1992, only Matsushita Electric, whose divisions have a high degree of autonomy, used the profit centered approach.

The major reason that most Japanese companies do not rely on the profit centered approach is that company data is considered to be highly confidential. Since the ability of the profit center manager to seek the lowest price for services is implicit in the profit centered approach, the company would have to be willing to outsource its information processing function. If the company cannot accept this, they must stick to a cost centered approach.

Advantages and Disadvantages of Chargeback Systems

If the chargeback system is introduced, costs for information processing can be precisely calculated, unnecessary information services can be eliminated, the cost/benefits of information processing can be analyzed, and optimal utilization of information resources can be obtained. For these reasons, using the chargeback system is not only advantageous to the information processing department but is also useful to the user departments.

Advantages of the Chargeback Systems

First, introducing the chargeback system can be a safeguard against providing unnecessary and unjustified services. If information processing services are charged to user departments, they will probably postpone or cancel unnecessary information processing. This should increase the effectiveness of resource use.

According to Choudhury's survey (1986), the main reason American companies give for using the chargeback system is to optimize resource utilization. The controller of Assist, a successful software house with an American president in Japan, believes that the chargeback system is indispensable to increasing the efficiency of departments that use data processing services. Of course, just using the chargeback system will not, in and of itself, eliminate requests for unnecessary and unjustified information processing. Cost/benefit analyses and internal auditing must also be used.

Second, increased awareness of cost control will lead to more careful consideration of the costs and benefits to the information processing and user departments. Let us assume that the information processing department is set up as a profit center. If the cost of information processing is lower in other divisions or at information processing centers outside the company, then that service can be requisitioned from other providers. Thus, the information processing department will have an incentive to provide good quality service at low cost. Since a premium price will be charged for urgent jobs and services, the departments requesting the services will need to carefully consider their requests.

Third, the chargeback system gives the user department a clear picture of the costs it is incurring. If the departments utilizing the information processing services do not have to bear the costs for data processing, then the true cost to that department may be unknown. As a result, decision making by managers may be based on incorrect costs and it is possible that this will lead to poor decisions.

The Disadvantages of the Chargeback System

There are also some important disadvantages to the chargeback system. The weak points are difficult to see during data entry; however its disadvantages become apparent when trying to apply it to the software development process.

First, there is the danger that decisions will be made from a short-term perspective. The advantages of investing in information system improvements cannot always be quantified, but there are many times when improved information processing is essential to improving operations. Because a chargeback system focuses solely on costs rather than continuous improvement, there are times when necessary improvements will not be made because of the high cost.

For example, Toyota once used the chargeback system but has recently discontinued it. Substantial information processing costs associated with improvement in old operations had made company managers skeptical about developing new systems. Thus, it indirectly suppressed administrative improvements with important long-term benefits. This concern that management would become short-term oriented is the major reason given by Japanese executives for not introducing a chargeback system

Second, if the chargeback system is introduced there will inevitably be conflict between the information processing department and the various user departments. The reason for this is the insolvable problem of transfer costs. No matter how precise, chargebacks are still allocated costs, not direct costs, with all the obvious inherent problems.

Third, clerical cost to maintain the chargeback system can be high. There are also hardware costs, software costs, and personnel costs required to run the chargeback system. For these reasons a chargeback system should only be introduced when the benefits clearly exceed the costs.

Designing a Chargeback System

When designing a chargeback system, three things must be determined in advance: the selection of the allocation basis, the types and range of costs to be transferred, and the level of sophistication of the system.

Single and Multiple Factor Methods

The first decision is what the basic principles of allocation will be. There are two methods of allocation, the single factor method and the multiple factor method.

The single factor method is a means of allocating cost to the user department using a single factor for allocation. This single factor method is reasonable when the information processing department is small and the same or similar information processing tasks are performed frequently. If it is to be a factor related to labor, the allocation basis might be the labor-hour or the labor cost. If it is to be a factor related to the machines, then the hour that the central processing unit is in use can be used. The hour that the CPU is operating is the most reasonable criterion when machine operation is the focus and there is only a little system development and maintenance. When basing it on the machine-hour, the following formula is in general use:

$$\text{Fee allocated} = \text{application rate} \times \text{CPU use}$$

The advantage to this method is its simplicity. On the other hand, if the allocation basis is not accurate, the user department will be dissatisfied, and conflict with the information processing department will increase. According to Nishizawa's survey (1991), the most frequently used allocation bases by information processing departments in Japan are: the amount of sales (24 percent), followed by the labor-hour of data processing (19 percent), and the machine-hour of CPU (13 percent). The sales amount is an unfortunate choice for the allocation basis, because hard work and increased sales will lead to higher costs. Not only is there no reason for this, but it will make the user department unhappy. Therefore, an allocation formula that is based on a user output, such as sales, should not be used.

The multiple factor method uses several allocation factors. The factors are computer resource variables, such as the CPU-hour, the number of lines printed, the volume of core memory used, disk and tape utilization, input/output transactions, and the volume of data in a directory. The recent progress in technology has made measuring these factors relatively easy. The multiple factor method is the most reasonable in large organizations with mainframe computers and minicomputers because cost assignment is based on specific resources that the user consumes.

However, whenever multiple factors are used, complexity and sophistication increase. Thus it may become more difficult for the user of the data processing services to understand and control (McGee 1986). For example, units which measure the resources of a computer, such as the number of lines printed or the volume of core memory used, are difficult for the user of data processing services to understand. Moreover, the user of the data processing services cannot effectively control these factors. Thus, when the multiple factor method is used, its complexity and sophistication may increase the frustration of the user (Finny 1981).

Types and Range of Costs to be Transferred

In the United States, the costs accumulated in the information processing department are in most cases based on the principle of full recovery of costs. In some cases they are based on partial recovery of costs. When the full recovery approach is used, all of the costs of the information processing services department are passed to the department requisitioning the services. As a result, there are no costs ultimately associated with the information processing services department itself. On the other hand, with the partial recovery method, a portion of the cost for information processing is passed to the user department and a portion of the cost remains with the information processing department.

Full recovery approach

The purpose of the full recovery approach is to eliminate the costs to the information processing department, i.e, to transfer all of the costs to the department using the services. The simplest way of doing this is to use cost accounting techniques, identify the services used, and assign them to the user department.

The defect in this method becomes apparent if there is idle equipment: the rates will increase because of idle capacity, which may interfere with use of the information processing system (MAP 1987).

Partial recovery approach

In the partial recovery approach, only a portion of the costs of information processing are recovered, so that the calculation is more complex than for the full recovery approach. There are two reasons why this approach is used.

First, it is grounded in the logic that only direct costs should be recovered, and that indirect costs should be borne by the information processing department. For example, in the case of the services of a programmer, the cost of the salary of that person would be transferred to the department using the services. However, costs related to the location such as depreciation of equipment, costs of consumable items, and overhead like utilities would remain in, and be borne by, the information processing department.

Second, it is also supported by the logic that although the cost of specific services should be charged out, the cost of other services should be carried by the information processing services department. According to Perry (1987), in a large insurance company on America's west coast, the costs for computer processing and data entry are passed to the department using the information processing services, but the costs of the system and the programming service are borne by the information processing department. Similarly, in a large Japanese steel plant, the costs of the information processing services are charged to the department that uses the services but the costs of system development are shouldered by the information processing services department.

Full recovery vs. partial recovery approach

The partial recovery approach is superior to the full recovery approach when the data processing equipment has idle time and the manager wants to make better use of the computers within the company. However, since the accounting for the partial recovery method is more complex, it is best to become fully acquainted with the full recovery approach first and then switch to the partial recovery approach.

According to a survey on American companies, the partial recovery method is used by only a small number of companies (Solomon et al. 1985). Forty percent of the companies surveyed use the full cost recovery method, 14 percent charge for specific tasks only (partial recovery), 12 percent do not include it in the budget since the information is for internal IS department purposes, 12 percent are charged when necessary, and 21 percent are not charged. According to McGee's 1987 survey, full cost recovery factors are used by 73 percent, direct computer costs plus overhead and the operator's salary are used by 10 percent, direct computer costs by 10 percent, and 7 percent use other methods.

Contrary to American practices, more Japanese companies use the partial recovery method than the full recovery method. For example, JIPDEC (1992) disclosed that out of 51 percent of the companies that assigned information processing costs to user departments, 30 percent assigned only partial recovery costs and 21 percent assigned full recovery costs to user departments.

From the point of view of responsibility accounting, the departments that use the information processing services should not have to bear the burden of any inefficiency in the information processing department. Thus, the use of standard cost is suggested whether using either full or partial recovery approach.

The Level of Sophistication of the System

When we speak about the chargeback system, there are various levels of sophistication. At the lowest level, it only allocates costs for the information processing department when necessary. At the highest level, it not only has close ties with the budgetary control system but also integrates chargebacks into financial statements. The level selected depends on the degree of systematization. We found three general levels from the results of surveys conducted during company visits and discussions with managers.

1. Costs allocated only when necessary for special studies or when information processing costs are allocated with other head office costs.

2. Costs for information services are controlled only by the budget system. These further break down into companies that practice variance analysis and companies that do not
3. Chargeback system is used for budgetary control, is completely systematized, and is included in the financial statements.

Allocation only

Some companies allocate the costs of the information processing department only for special studies, but it is difficult for the chargeback system to work efficiently in this circumstance. In other companies, cost of information processing department is included in headquarter costs and is allocated to other divisions but not integrated with the budgetary control system. Though these companies may be said to be allocating the costs, this is not a chargeback system in the true sense. According to the survey done by Solomon and Tsay (1985), 12 percent of the companies used this method in the United States. My survey shows that 12 percent of the Japanese companies surveyed (146 companies) also used this method.

Allocation and budgetary control

Some companies control the costs of the information processing department within the budget system but these have no link to the final settlement of accounts of financial statements. From the data available, this type of system is more often seen in Japanese companies. The author's survey tells us that one-quarter of companies surveyed use this integrated system, while only 15 percent of the companies surveyed used the chargeback system with budgetary controls that are not tied to the financial statements.

On the other hand, there were many Japanese companies that either did not perform any variance analysis, or even if they did, did not link the results to an evaluation of performance.

Regardless of how much they might try to keep costs down in the information processing departments and restrict the usage of information processing services, they cannot control the very occurrence of costs. Trying to restrict the use of information services to reduce costs is not good for the future of the company, just as reducing R&D expenditure is not good.

When allocating on the basis of actual results, there is the danger that the IS department will be pushed by the user departments to the point of inefficiency. Thus, Japanese companies with a high cost-consciousness try to avoid this danger by using standard costs for unit prices even though the actual use will change the cost per unit.

In addition, since an increase in the volume of use is often for the purpose of improving operations, it is not always a bad thing for a company that its information processing costs are high. Of course there should be no waste in information processing. For example, logistics (distribution) information processing costs should be minimized as Naka (1989) maintained. In such cases, variance analysis is preferable. Nonetheless, the money for information processing, like that for R&D, is often spent for the future of the company. In contrast to American companies, Japanese companies do not generally link the results of variance analysis and performance evaluation in these cases.

Allocation, budgetary control, and financial statements

In the most complete system, the results of the chargeback system are not only used in allocating information processing department costs and in budgeting, but are also included in the settlement of final accounts. It is possible to extend this system into every nook and cranny of a company. This type of system seems to be quite common in the United States. Many companies with integrated systems carry out variance analysis at the end of each accounting period. A survey by McGee (1987) revealed that 92 percent of American companies perform variance analysis.

The Procedures of the Chargeback System

There are no set procedures for introducing the chargeback system. The following description is based on comprehensive budgetary control using on the multiple factor method, and is grounded in the full recovery approach. Any of the methods introduced below can be abbreviated or modified to suit the size and type of the company and the degree of sophistication of its information processing.

Budgeting for the Information Processing Department

Since the purpose of the chargeback system is the recovery of information processing costs, it is necessary to predetermine costs in the annual budget. The budget of the information processing department may be prepared together with the comprehensive budget. However, when it is prepared separately from the comprehensive budget it is more effective for implementation.

The Resources and Their Measurement

The resources that will be used as the basis for allocation and the measurement units should be defined in advance. In both the United States and Japan, the machine-hours that the CPU is used is typically selected as the basis when using a single factor. When using the multiple factor method, what is to be measured should be chosen after considering the nature and performance of the company. Typical resources and their units are shown in Table 10-3.

Table 10-3. Resources and Measurement Units

Resources	Measurement Units
Use of CPU	Hours of CPU
Use of disk	Output/input
Library	Volume of data
Use of tape	Output/input
Print	Lines, pages
Data entry	Labor-hours of operators

Estimating Normal Operations

Determining the normal operating level of resources is essential for determining rates. Basing the rates on a standard value protects the user departments from bearing the burden of any inefficiencies in the IS department. A standard that may be realistically expected (referred to as the expected actual standard) is preferable to using maximum capacity as the standard. Some examples of the results of determining usage levels are shown in Table 10-4.

Table 10-4. Types of Resources and Use Levels

Type of Resources	Activity Level of Resources
CPU	1,600 hours
Output/input of data	520 × 10
Output/input of disk	400 × 10
Lines of printing	300 × 10
Operator hours of data entry	24,000 hours
Hours of programming and analysis	29,000 hours

Budgeting and Allocating

An allocation for each service is done as is shown on the left side of Table 10-4. Sometimes, several resources are consolidated and allocated.

Application of Resources

The most direct computation method is to divide the amount in the cost pool by the use level of the specific resource. For example, if the budgeted amount for the CPU is ¥ 32,000,000 and the usage level for the CPU from Table 10-4 is 1600 hours, the application rate is computed with the formula shown on the next page. The application rates for each of the other resources are computed similarly.

$$\text{CPU rate} = \frac{\yen\,32{,}000{,}000}{1600\text{ hours}} = \yen\,2{,}000/\text{hour}$$

Computing the Multiple vs. Single Factor Method

With the multiple factor method of allocation, the amount to be allocated is computed with the rate determined for each resource. Then, an itemized invoice is sent to each of the user departments. The advantage is that the user departments can check the amount consumed. However, many user departments do not need continual invoicing and are satisfied if the invoice is simply available should they need it.

There is also a method of calculating the application rate for each output unit rather than for each resource consumed. This would be based on the actual output; for example, the number of documents sent, the number of persons whose wages were computed, or the number of invoices generated.

When using the single factor method of allocation, the application rate for each unit is calculated by dividing the average cost of the resource by the number of units output. The rationale for using the single factor approach is its simplicity. Proponents argue that it often leads to the same results as the multiple factor method and that the amount is not an issue, the process and the method of assignment are more important. The output unit should be carefully defined. Data for many months must be collected to determine the relationship between the amount of a resource consumed and output.

All of these procedures will be more accurate when the output volume is large. However, as usage increases, the problem of overapplied overhead arises. To help compensate for this, the standard price should be recalculated at the beginning of each period. Regardless of whether it is used in evaluating performance, variance analysis should probably be done at the end of each period to reevaluate the allocation rate.

Linking Budgetary Control to the Financial Statements

In the settlement of accounts, multiplying the standard price by the actual amount produces the closing figures. There will be a difference in cost between standard and actual costs. The cost differences will be charged off all at once at the end of the accounting period.

Conclusion

The chargeback system creates cost awareness in the user departments and it is one of the most appropriate methods for evaluating performance. It is also an effective method for controlling operating and maintenance costs for computer software. It is much more popular in American firms than in Japan, probably because it is a more effective tool for performance evaluation than for strategic cost management.

For example, with the chargeback system, the substantial information processing costs associated with improvement in old and inefficient operations may make managers skeptical about developing new systems. The needs of departments that use information processing services but are not amply endowed with funds may be overlooked. It also does not take into account how much benefit or profit the department using the information processing service derives from that service. Thus, it does not guarantee that a company gets the information service that it really needs.

There is an alternative approach for controlling the costs of the information processing department: Treat the costs of software development as a capital investment in the same manner as investment in plant and equipment. As Seed (1984) rightly commented, dollars invested in software are essentially no different than dollars invested in hardware. Because of the onetime nature of the expenditures and the substantial stakes that may be involved, it is often desirable to include software development as part of the capital expenditure budget for financial planning purposes.

Last, but not least important, are the problems raised with the recent onset of downsizing and end-user computing. The development of end-user oriented computing has made implementation of chargeback systems more difficult (Hashimoto 1993). An example may best explain the situation. In early 1992 Matsushita established a new total information system using Macintosh personal computers and artificial intelligence (AI) software. This system was developed by the initiative of Information Processing Kaizen Department in Matsushita. However, the system was completed only with the cooperation of a large number of staff from the user departments. In this case the chargeback system discussed here could not be applied because information processing cost does not incur at one place—therefore it is impossible to accumulate the cost.

In the new environment of open systems, a new cost management system for information technology must be constructed in addition to the chargeback system discussed in this chapter. Since September 1994, the author has chaired the Committee on Cost Management Systems under an Open System with MITI financial support. One of the Committee results for the first year is the publication of "A System Justification Methodology under an Open System" (Sakurai 1995). This methodology will clarify decision-making when a company is contemplating introducing information technology.

The new system has three characteristics. First, it can measure strategic benefits as well as operational profit gained from introducing new information technology. Second, it can measure the effects given to information service departments in the downsized environment. Third, it can measure the benefits that the user departments can gain from the new information technology. When a large number of personal computers and workstations are used in a client/server system, those effects can be measured in yen or dollars.

References and Further Readings

Choudhury, Nandan, Sumit Sircar, and K. Venkata Rao. 1986. "Chargeout of Information Services." *Journal of Systems Management* (September) pp. 16, 21.

Finny, John E. 1981. "Controlling EDP Costs, Assisting Clients in Keeping Computer-Related Costs Down." *Journal of Accountancy* (April) p. 54.

Fremgen, James M., and Shu S. Liao. 1981. *The Allocation of Corporate Indirect Costs.* National Association of Accountants, p. 37.

Hashimoto, Giichi. 1993. *Management of Information Systems*, Hakuto Shobo Publishing Company, p. 89.

JIPDEC (Japan Information Processing Development Center). 1992. *White Paper of Information,* pp. 61-62.

McGee, Robert W. 1986. *Accounting for Data Processing Costs.* Quorum Books, pp. 1-2, 5.

McGee, Robert W. 1987. "Accounting for Data Processing Costs." *Journal of Accounting and EDP*, p. 44.

MAP Committee, SMA 4F. 1987. *Allocation of Information Systems Costs* (Allocation Method 18; Allocation Bases 20; Responsibility Accounting Consideration 13).

Naka, Mitsumasa. 1989. "Cost Accounting for Logistics Information Costs." *Journal of Business* (School of Business at Asahi University), pp. 87-99.

Nishizawa, Osamu. 1991. "Allocation of Costs of the Information Processing Department to the User Departments." *The Waseda Commercial Review* no. 344 (March) p. 64.

Perry, William E. 1988. "Implementing DP User Chargeback Systems." *Journal of Accounting and EDP* Vol. 4, no. 3(fall) pp. 9-10, 13-16.

Sakurai, Michiharu, et al. 1987. *Cost Accounting for Software.* Hakuto Shobo Publishing Company, p. 176. In this first edition, the issues concerning cost accounting for software to be sold are covered.

Sakurai, Michiharu. 1990. "Survey on Accounting Practice for Software." *Accounting* (July) pp. 92-93.

Sakurai, Michiharu. 1995. "System Justification Methodology under Open System: Measure Strategic Benefits through New Technique." *Nikkei Computer* (May 15) pp. 129-132.

Seed III, Allen H. 1984. "Management Accounting for Software Development." *Corporate Accounting* p.15.

Solomon, Lanny, and Jeffrey Tsay. 1985. "Pricing of Computer Services: A Survey of Industry Practices." *Cost and Management* (March-April) p. 6.

CHAPTER 11

Performance Evaluation for High Technology Companies

As a result of the active introduction of American management accounting theory and practice into Japan after World War II, the gap between management accounting theory and practice in the two countries has narrowed noticeably. At the same time, a number of important differences in Japanese and American management accounting have developed. These are due both to cultural differences and to the development, since the early 1970s, of unique Japanese management accounting concepts and techniques that arose especially from the introduction of FA.

One of these conspicuous differences is in the area of performace evaluation and measures. In the United States, great emphasis is placed on return on investment (ROI). In contrast, Japanese management generally places emphasis on periodic income. Prior to the oil crisis of 1973 that tendency was particularly striking in Japan. Since the 1980s, high technology companies have shifted to using the return on sales (ROS) method. Very recently there has been a gradual shift toward using ROI, as Japan's leading economists point to the importance of effective resources management.

Frequent Use of ROI in American Companies

The ROI has been used by typical American companies since it was originated by Du Pont in the 1920s. Several surveys show that this use continues. For example, in the 1966 survey by Mauriel and Anthony 52 percent of responding companies were using only ROI as a measure of performance, 42 percent of them were using both ROI and residual income (RI), and only 6 percent of the companies were using RI alone.

In the debate over ROI or RI as a measure of the performance of divisions, such academics as Solomons (1965), Mauriel and Anthony (1966), Dearden (1969), and Reece and Cool (1978) argued that ROI is inferior to RI as an overall means of measuring financial performance. There are also arguments for using multiple performance measures (Berliner & Brimson 1988; Johnson 1990; Dixon, Nanni, and Vollmann 1990). Lander and Bayou (1992) suggested using modified ROI in robotic factories. Despite these arguments, the use of ROI has not diminished and in fact appears to be growing. For example, the Reece and Cool survey published in 1978 reported that ROI was being used more than in the past. More specifically, according to that survey, 65 percent of all the companies were using only ROI, 28 percent were using both ROI and RI and only 2 percent were using RI alone for measuring performance. The work of Miller (1982), who reviewed some of the material concerning methods of evaluating performance over the previous 25 years, demonstrates that interest by American companies in using ROI as a means of measuring performance is increasing.

The term "investment center" is gradually coming to be used synonymously with the measurement of ROI. Most surveys noted above were undertaken to evaluate the divisional performance of investment centers. We could therefore safely generalize that ROI is typically used by companies having divisional organizations. Nonetheless, even with the results of the surveys noted above, it is still surprising that ROI maintains its popularity for divisional performance measurement. Lander and Bayou (1992) described it as follows, "In spite of criticisms, ROI enjoys external popularity in evaluating managerial performance." They gave three reasons: (1) ROI is simple to understand. (2) It combines three critical performance measure variables—sales, earnings, and investment. (3) It is popular with financial analysts, investors, creditors, and other external information users.

Periodic Income, Multiple Measures in Japan

In contrast, Japanese managers normally use measures besides measures of profits for evaluating divisional performance. According to the survey conducted by Tsumagari and Matsumoto (1972), only 32 percent of the 307 companies surveyed were using profit measures to measure divisional performance. Other performance measures were sales growth, productivity, and market share.

The number of companies organized as divisions in Japan is relatively small in comparison with the United States. Instead, there are many subcontractors and other affiliated companies in Japan. This is one of the unique characteristics of Japanese organizations. Consequently, when comparing performance measurement systems we need to examine not only divisions but also affiliated companies. The surveys conducted by Miyamoto and Matsutani (1982) found that multiple measures besides profits are often used to measure affiliated company performance. In a study conducted by Aoki (1975), 49 percent (66 of 136 companies) replied that they were using periodic income to measure profit. In contrast, only 16 percent (23 companies) were using ROI. This study indicated that more companies used volume of production (32 percent) or sales growth (24 percent) as measures of performance than were using ROI. These survey results represent typical approaches to performance evaluation in Japanese companies.

Why do many Japanese managers use periodic income for evaluating divisions and affiliated companies? We will discuss the following three reasons.

1. The backwardness of the managerial consciousness of Japanese managers and of the management system.
2. Institutional differences such as the weak power of the stockholders and lack of divisional autonomy
3. The high rate of economic growth and the financing alternatives

Japan's Backwardness

The first thing that the reader may think of after seeing the results of these surveys is that the awareness of Japanese managers and the management system itself may be backward compared to the United States. That is because it is generally believed that the relationship between sales profitability and turnover is theoretically a better way to measure profits than using the absolute amount of profits (Symonds 1975). To put it another way, measuring the efficiency of capital by ROI is believed to be superior to measuring the absolute amount of profit. Since ROI is not used very much in Japan for measuring divisional performance, the understanding of many Japanese managers and the management system itself may be looked upon as backward.

More than 15 years ago a panel discussion was held at the annual meeting of the Japan Accounting Association held at Kobe University. In the course of that discussion it was stated that periodic income is the primary measure used by Japanese companies to evaluate divisional performance because it can effectively measure "volume expansion." One of the participants said, "The consciousness of those managers is very backward." His response had been the typical attitude of Japanese academics toward practitioners until the 1970s.

This may have been true for Japanese managers during the 1950s and early 1960s when Japanese management systems had not yet been perfected. And indeed, the management system in typical Japanese companies is still behind that of the United States. However, it may be a mistake to conclude that ROI techniques are not used simply because of a lack of sophistication, for the following reasons.

1. ROI is easy to comprehend. So, it is inconceivable that Japanese top management cannot master ROI.
2. ROI is not necessarily superior to the amount of periodic income. In fact, conceptually, RI is superior to ROI.
3. Though Japan's management systems are not as well developed as those in the United States, it may safely be said that they have largely perfected the use of such management control systems as budgets (Sakurai 1983).

In short, although the backwardness of Japanese managers might have been the reason for some differences during the 1950s and early 1960s, this same reason cannot fully explain why ROI is not often used today for evaluating divisional performance.

Institutional Differences

Up to this point we have not discussed why multiple measures have often been used in evaluating the performance of Japanese divisions and affiliated companies. It is important here, since this issue is also related to institutional differences. This subject relates to General Electric's RI and other key goals in comparison with the ROI of Du Pont.

Quite a few academics argue for multiple performance measures. They say that if Japanese managers were to use a single measure for measuring divisional performance, such as ROI, they would be using an unrealistic, though simple, economic model. Others would argue that while the ultimate goal of American companies may be profit, Japanese companies not only seek to achieve economic goals but also place more emphasis on the welfare of their employees. As Kono (1978) points out, a major reason for this difference may be the difference in the power held by the stockholders. In the United States, stockholders are powerful and thus, earnings per share and ROI are emphasized. On the other hand, in Japan the ratio of equity capital is low and therefore the power of the stockholders is weak and short-term goals are not emphasized.

Furthermore, there is a tendency to take a long-term view of manager performance. This tendency is related to the lifetime employment tradition. It also means that Japanese companies tend to evaluate the performance of employees more in terms of the long-run rather than short-run results. When one considers that the simple measure of profits carries with it the danger of emphasizing short-term results at the expense of long-term flexibility and income (Richard 1978), one can understand the reason why a single measure of performance is not very often used in Japan.

Then, why is ROI not emphasized as a measure for evaluating divisional performance in Japan? In terms of institutional factors, some may suspect that it is related to the relative strength of the autonomy of divisions. That is to say, historically, companies in the United States have often made what was originally a factory of the company into an independent division. And they very often make companies that have been merged and acquired into operating divisions. In Japan, however, mergers and acquisitions (M&A) were rare up until the early 1980s. Where divisionalization is used by Japanese companies, the power of the corporate headquarters is strong and the autonomy of the division weak. Thus, one would suspect that the measure for evaluating divisional performance would end up being measures in short-term perspectives such as sales volume or the amount of periodic income.

However, the reason why ROI is not emphasiszed in Japanese divisions is not only weak divisional autonomy, because ROI is not considered important even in companies without the divisional organization structure. The reason Japanese divisions do not use ROI cannot be solely explained by the fact that divisions are not completely autonomous, as was clear from the survey by Tsumagari and Matsumoto (1972).

The characteristics of the Japanese management style that explain the low use of ROI become clearer when one takes a look at corporate goals instead of performance evaluation. According to the survey made by the Bureau of the Economic Planning Agency (1976), the most common corporate profit goal was periodic income (especially ordinary profit) (70 percent), followed by ROI (16 percent) and ROS (13 percent). It was clear from this survey that periodic income was often used instead of ROI even as a corporate goal in Japan.

But the following contrary view may be advanced. In the United States, ROI may be used less for setting goals than for evaluationg performance. However, in the study by Reece and Cool (1978) previously referred to, 70 percent of the companies that use ROI as a means for evaluating performance also adopt ROI as a corporate goal. In addition, it is clear from the Kagono survey (1983) that far fewer Japanese companies set ROI as a corporate goal than do American companies. The author's survey (Sakurai 1992) also supports Kagono's conclusion. In terms of performance evaluation, periodic income was used by 82 percent (103 companies), ROI was used by 10 percent, and ROS was used by 26 percent (33 companies). In terms of profit goal, periodic income was used by 88 percent (126 companies), ROI was used by 7 percent (10 companies), and ROS was used by 20 percent (28 companies). The 573 companies surveyed were listed on the Tokyo Stock Exchange and included five assembly-oriented industries. Out of those companies, 142 gave multiple answers to the question of profit goal and 125 to the question of performance evaluation.

It is interesting that Japanese managers are apt to place greater importance on ROI for setting long-range goals. For example, ROI is frequently used in formulating long-range business planning. According to the surveys by the Enterprise Management Association (1967, 1974), ROI is most important in long-range business planning though it is, as in other cases, not very important among various indexes. This means that typical Japanese managers manage their companies with ROI as one of the most important long-range corporate goals though they make little use of ROI for short-term business operations.

In sum, the presumption emerges that ROI is not often used by Japanese companies because the autonomy of the division is weaker than in the United States. However, this is not necessarily the entire reason for the lower use of ROI in Japan. The argument about the weakness of the stockholder is persuasive and may be the most important reason why ROI is not often used. However, despite the fact that the difference in the relative strength of the stockholders has become an explanation for why EPS (earning per share) is used less in Japan, it is not sufficient alone to explain why ROI is not given more emphasis. There must, no doubt, be some other reason.

Rapid Economic Growth, Inflation, and the Financial Methods

During the period of rapid economic growth of the latter half of the 1950s to the early 1970s, Japanese companies expanded the scale of their operations. For capital, they relied chiefly on borrowing from banks. Tax laws in Japan, like those in the United States, treated the interest on borrowed money as a cost, so that the lowest cost means of financing, in terms of capital costs, was to borrow from the bank rather than to increase equity and pay dividends.

During this period of rapid economic growth, it was not difficult to make a substantial profit. Japanese companies borrowed large amounts and the interest was not much of a burden because of the high rate of inflation. Generally, the danger of bankruptcy was relatively small, and borrowing from a bank was easy since the land used as collateral for the loans was mortgaged with extremely reliable mortgaging procedures. The general public did not usually invest in stock, other than through institutional investors. Since consumer bank deposits were large, banks were always able to provide ample amounts of capital.

Because Japan had these means of providing capital, the ratio of net worth to capital was very low. It is still low, moving to around 27 percent in 1993. The net-worth-to-capital ratio of firms capitalized at more than ¥1 billion was the lowest, 17.7

percent, in 1980, but it moved up to 25.1 percent in 1987 and by 1991 it was 26.7 percent. Japanese companies were able to apply the leverage effect by using bank loans for new investment. Incidentally, although the rate of earnings on total capital is less than in American companies, the ratio of net profit to net worth surpassed that for the United States during the early 1970s (Economic Planning Agency 1976).

After the war, particularly due to the high rate of inflation that continued until the late 1970s, Japanese managers felt keenly that the best way to insure the continued existence of their companies was rapid expansion of production facilities. Managers learned through experience that by investing in land, equipment, and machinery they could obtain significant capital gains (Keizai Koho Center 1982). This also lead many companies to borrow aggressively from banks at low interest. Moreover, the cost of obtaining capital through bank loans was relatively low compared with the United States (Bank of Japan 1982) although there is some disagreement about how low the cost actually was (Kester and Luehrman 1992).

In these high-growth economic conditions and with these financing methods, it was natural to invest aggressively in order to increase sales volume even with new investments which may exert an adverse effect on short-term profitability or ROI. During the time of rapid economic expansion, those companies that grew slowly even with high ROI disappeared sooner or later. Thus, for many Japanese companies the best corporate goal was to expand sales volume, market share, and income. As for profits, most companies aimed to increase the amount of profits after deducting interest, in other words, ordinary income.

ROS for High Technology Companies

We have discussed whether it is better to use ROI or periodic income as the profit goal or measure for evaluating divisional performance. However, we also need to consider the role of ROS.

Problems with ROS

The argument against ROS is clear. As we see in the formula below, the turnover or efficiency of capital is not included in the ROS formula, so that ROS is an incomplete measure of performance for overall profitability or evaluation.

$$\text{Return on Investment} = \text{Return on Sales} \times \text{Turnover}$$

$$\frac{\text{Income}}{\text{Investment}} = \frac{\text{Income}}{\text{Sales}} \times \frac{\text{Sales}}{\text{Investment}}$$

Problems with ROI

Since the emphasis on ROI brought about a tendency to avoid positive investment in the United States, American academics are generally critical of using ROI. They have become critical of ROI not only because attempting to increase ROI in order to maintain short-term profits to stockholders inhibits investment in plants and equipment, but also because it tends to restrict investment in high technology R&D expenditure. Nonetheless, in practice, typical American companies emphasize ROI. Various surveys reveal that ROI use is growing year by year.

What does happen when companies place emphasis on ROI in their operations? Let's explain with an example. Division A is presently reaping a profit of ¥6 billion on an investment of ¥30 billion. An investment proposal to get into semiconductors is presented. The amount to be invested is another ¥30 billion. The outlook for profits and competitive superiority will be achieved in the future. But immediately following this investment there will be no profit for three years. Thus, in the three years immediately following the initial investment, the ROI will drop from 20 percent to 10 percent as is shown in Table 11-1.

As a result of this investment, the stockholder must be content with low dividend payments for at least three years. Thus, stockholders who want more dividends would oppose this proposal and this strategic investment might not be attempted. Similarly, if

Table 11-1. ROI Before and After Investment

Alternatives	Income / Investment	=	ROI
Status quo	¥ 6 billion / ¥ 30 billion	=	20%
After investment	¥ 6 billion / ¥ 60 billion	=	10%

investment in CIM is canceled because short-term direct profits cannot be obtained, Division A would probably not be able to maintain its competitive superiority over the long-run.

In Japan, the academics in business schools strongly advocated placing emphasis on ROI (like American managers had been doing), but in practice emphasis was placed on periodic income not ROI. Put another way, companies have been managed with little concern for ROI, at least up until 1970s.

This was probably the consequence of the combination of a lack of pressure to emphasize ROI (because of the weak position of the Japanese stockholder) and the healthy capital gains from inflation and rapid economic growth. Management emphasized periodic income for these reasons.

However, in the 1980s economic features of the business environment changed dramatically, and the rate of capital gains slowed dramatically. The structure of manufacturing continues to evolve rapidly from large scale heavy industry to lighter, smaller scale manufacturing employing the highest and most advanced technology. The life cycle of consumer products has been radically shortened as has the life cycle of manufacturing equipment. Although the use of periodic income may have been appropriate for measuring the performance of large scale heavy industry, it may not be an appropriate measure for evaluating the performance of advanced industry where technological innovation is constant.

Reasons for Using ROS

In the United States, the position held by Riggs (1983) that "it is important to maximize ROI" holds sway in advanced manufacturing technology. However, major Japanese companies advanced in high technology place greater emphasis on ROS for both profit planning and the evaluation of performance. For example, of 32 companies the author visited in 1984 that had proceeded with FA, 72 percent (23 companies) were making use of ROS. They had progressed much farther than the five companies that were using periodic income and the four companies that were trying to use ROI. Is this simply a coincidence? Are the companies that emphasize ROS being managed without concern for the efficiency of capital?

While the companies that emphasize ROS do not all do so for the same reason, the following are given by these companies as their reasons for using ROS.

1. Using ROS when a variety of products are produced makes the profitability of each product clear.
2. When a company is using target costing, it is easier to set prices for a variety of products. In the process of setting target costs it is convenient to designate ROS. For example, if ¥ 4 million's worth of automobiles are ordered and the target profit is 20 percent, then the target profit will be calculated as ¥800,000 (4,000,000 x 0.2).
3. It is nearly impossible, from a cost benefit perspective, to justify attempting to compute the ROI on each product when producing a variety of products with low volume.
4. Because the amount of capital is likely to vary in high technology industries, ROI cannot be reasonably calculated.
5. Large investments of capital are generally needed in order to produce profitable products in high technology industries. When one relies on ROI, however, desirable investments undertaken for the future are likely to be shown as unprofitable in the present.
6. If there is stable demand and no apprehension about declines in sales, ROI may be effectively employed. However, these conditions do not exist for high technology products.
7. There is little pressure from stockholders in such private companies as Epson.

Hitachi was known to consider ROI an important goal. But since 1979, it has evaluated performance according to "U rank," that is ROS, with the Du Pont performance evaluation index:

$$T\ (ROI) = U\ (ROS) \times K\ (turnover)$$

By evaluating divisions or subsidiaries with the U rank (ROS), Hitachi could easily measure the profitability and competitiveness of each factory against the others. It is also convenient because ROS is easy to calculate. In the corporate headquarters, on the other hand, "T" (ROI), which depends on the overall index, is emphasized.

Though we speak of companies that make ROS the target, that does not mean that they are managing their companies ignoring the efficiency of capital. For example, such companies as Toyota and Matsushita are operating to increase their efficiency of capital in a format that is separate from ROS.

Strategic Use of ROS

On the debit side of the balance sheet, assets are classified roughly as (1) cash and accounts receivable, (2) inventories, and (3) plants and equipment. In contrast to American business practices, Japanese companies cannot expedite the recovery of accounts receivable without the risk of losing customers because terms of payment are customarily predetermined. Reducing investment in plants and equipment in order to pay high current dividends may be advantageous to the stockholder, but doing this robs the company of its future. Thus, the critical asset among the three is inventories—and the goal is to reduce them.

Toyota is not only making use of JIT but is also using ROS for target profits. Why did Toyota originate the JIT system? Both JIT and target costing are intimately related to using ROS as a target profit. Since the only assets suitable for reduction in the Japanese situation are inventories, it becomes indispensable that inventories be reduced when using ROS. The use of ROS with target costing makes pricing easy and practical.

Thus, we see that Toyota's JIT is a policy designed to increase the efficiency of capital through reducing inventories, a specific form of assets. Similarly, Matsushita's Internal Capital/Internal Interest System and standard balance sheet are also means for increasing the efficiency of capital. The internal interest system is used for managing divisions. Internal interest is computed by multiplying internal capital by internal interest rate. A fixed rate of interest is levied on employed capital in order to establish in the division employees a consciousness of the importance of employed capital. The company's internal capital system is designed to make each division operate as if it were an independent company by allocating capital costs to each division. Both systems were originated at Matsushita. It is close to residual income. Some other companies call these the internal company interest system and the internal company capital system, respectively. Matsushita manages the company by placing emphasis on sales profitability because of the need for a management control system that can increase the efficiency of capital.

In this way, the peculiar defects of ROI can be eliminated by separating the ROI into ROS and the turnover. These defects are, as stated earlier, a negative effect on new investment in plants and equipment in some circumstances and the lack of clarity in the definition of capital for the ROI denominator. When thinking in this fashion, ROS may become an effective means of management in the age of CIM if it is used prudently.

Conclusion

There are two opposing views concerning the management approach in Japanese companies, which tends to ignore ROI in performance evaluation and the setting of corporate goals. One is the argument for greater use of ROI. Many business critics argue that failure to use ROI leads to extreme (and economically unsound) competition, and should be corrected. On the other hand there is the view that the lack of concern for ROI reflects a unique Japanese management style.

Needless to say, companies need high ROI in the long run. Nonetheless, some companies in the United States that have been managed in this way have lost markets. We should not ignore this historical lesson. As Mechlin and Berg (1980) have pointed out , using ROI in the United States has retarded investment in R&D enough to be called a "restriction on innovation." If Japan had not invested in new markets, had not expanded its facilities, or had failed to renew its equipment because it would require substantial R&D costs and hence a lowering of ROI, it would not be experiencing the prosperity of today.

We do not mean to argue that Japanese management has been absolutely correct up to this point. In one respect, the Japanese tendency to ignore ROI has intensified domestic and international competition among companies. In addition, as economic growth in Japan slows and inflation lessens, the attitude of managers with respect to capital investment has become conservative and gradually methods of obtaining capital have expanded to include numerous sources of funds, mainly equity capital (Oka, 1989). The author predicted that Japanese managers would become more conscious of ROI than in the past.

However, in the high technology companies since the oil crisis, ROS rather than ROI has been increasingly used for target profit. They began to control ROI by separating ROS and turnover. As a means of increasing the efficiency of capital, Toyota attempted to raise the turnover of inventory using the JIT system. Matsushita originated internal capital and internal interest systems as measures to increase the efficiency of capital. This is a management innovation. Having considered the above, we would argue that the ROS originated by Toyota (or Matsushita) should be considered a management innovation along with the well-known ROI of Du Pont, and the RI of General Electric. The relationship between these is shown in Table 11-2.

Since the collapse of the bubble economy in 1991, the Nikkei has emphasized shareholder wealth, and thus argued for increasing ROE. Recent Nikkei (1994) surveys revealed that both Mitsubishi, the nation's largest trader, and Nomura Securities,

Table 11-2. Three Indices for Performance Evaluation

Indices	Abbreviation	Major Companies
Return on investment	ROI	Du Pont
Residual income	RI	GE
Return on sales	ROS	Toyota

Japan's largest securities company, began to use ROE as a corporate goal for efficient operation of capital rather than pursuing the increase of sales volume, market share, or ROS. Currently, profit or volume oriented corporate goals alone do not dominate. Rather, a major goal of typical Japanese companies is pursuing effective management. Japanese managers wish to use corporate resources effectively. Thus, Japanese companies may gradually switch from using periodic income or ROS to using ROI or ROE.

References and Further Readings

Aoki, Shigeo, ed. 1975. *Management and Accounting of Affiliated Companies*. Japan Tax Research Association Publications, p. 259.

Bank of Japan. 1982. *Comparative International Statistics*.

Berliner, Callie, and James A. Brimson. 1988. *Cost Management for Today's Advanced Manufacturing*. Harvard Business School Press, pp. 3-6, 162.

Dearden, John. 1969. "The Case against ROI Control." *Harvard Business Review* (May-June) p. 126. Dearden implies that the causes called technical limitations, in this book the implementation limitations related to the decrease in R&D in the United States, make the application of ROI into Japan meaningless because of the high inflation there.

Dixon, Robb, Alfred J. Nanni, and Thomas E. Vollmann. 1990. *The New Performance Challenge: Measuring Operations for World-Class Competition*. The Dow-Jones, pp. 53, 112, 142-143.

Economic Planning Agency. 1976. *White Paper on the Economy for 1976: For the Foundation of a New Expansion*. Finance Ministry Printing Office, p. 157.

Economic Planning Agency (Survey Office). 1976. *Survey on Company Search Activities* (January).

Enterprise Management Association. 1967 and 1974. *Management Planning Survey*. Business Management Association (NAA Tokyo Chapter).

Johnson, H. Thomas. 1990. "Performance Measurement for Competitive Excellence." In Robert S. Kaplan, ed. *Measures for Manufacturing Excellence*. Harvard Business School Press, p. 65.

Kagono, Tadao, Nonaka Ikujiro, Sakakibara Kiyonori, and Okumura Teruhiro. 1983. *A Comparison of Japanese and American Companies: Theories on Strategies of Adapting to the Environment*. Nihon Keizai Shinbunsha Inc., p. 25.

Keizai Koho Center Japan. 1982. *An International Comparison, 1982*. p. 67. As a source comparing Japan and the United States, this is handy reference material.

Kester, W. Carl, and Timothy A. Luehrman. 1992. "The Myth of Japan's Low-Cost Capital." *Harvard Business Review* (May-June) pp.130-138.

Kono, Toyohiro. 1978. *Examples of the Newest Long-Term Operational Plans*. Doubunkan, p. 43.

Lander, H. Gerald, and Mohamed E. Bayou. 1992. "Does ROI Apply to Robotic Factories." *Management Accounting* (May) pp. 49-50.

Mauriel, John J., and Robert N. Anthony. 1966. "Misevaluation of Investment Center Performance." *Harvard Business Review* (March/April) pp. 98-105.

Mechlin, George F., and Daniel Berg. 1980. "Evaluation Research: ROI Is Not Enough." *Harvard Business Review* (September-October) p. 94. Refer to Hayes, Robert H., and William J. Abernathy. 1980. "Managing Our Way to Economic Decline." *Harvard Business Review* (July-August) p. 67 for a more detailed treatment.

Miller, Elwood L. 1982. *Responsibility Accounting and Performance Evaluation*. Van Nostrand Reinhold Company, pp. 100-129. There is a splendid introduction to Miller's work by Professor Koga Tsutomu (See Tsutomu, Koga. "A Test of Residual Profits and Performance Reports." Studies in Commerce (Fukuoka University) Vol. 27, nos. 1 and 2, pp. 257-297).

Ministry of International Trade and Industry. 1983. *1982 Edition New Management Index*. Finance Ministry Printing Office (April 11) p. 111.

Miyamoto, Masaaki, and Matsutani Yasuji. 1982. "Research on Management Systems in Affiliated Companies." *Accounting* (November) pp. 95-96.

Nikkei. 1994. "Corporate Goal—More Return than Volume." *Nikkei*. Nihon Keizai Shimbunsha Inc. (June 3).

Oka, Masao. 1989. "Japanese Company Finance in Transition." *Accounting* Vol. 41, no. 11 (November) pp. 16-21.

Reece, James S., and William R. Cool. 1978. "Measuring Investment Center Performance." *Harvard Business Review* (May-June) pp. 29-30, 42.

Richards, Max D. 1978. *Organizational Goal Structures.* West Publishing Company, p. 10.

Riggs, Henry E. 1983. *Managing High-Technology Companies.* Life-time Learning, p. 202.

Sakurai, Michiharu. 1983. "A Comparison between Japan and the United States in Management Accounting Practice." *Business Review of Senshu University* (Memorial edition commemorating the retirement of Professor Yamada Ichiro) no. 36 (July) pp. 117-137.

Sakurai, Michiharu. 1992. "Japanese Management Accounting Practices: Analysis of Mail Survey for CIM." *Business Review of Senshu University* no. 55 (October) p. 134.

Solomons, David. 1965. *Divisional Performance: Measurement and Control.* Financial Executive Research Foundation, p. 64.

Symonds, Curtis W. 1975. *Profit Dollars & Earnings.* New York: A Division of American Management Association, New York, p. 75. Here he comments on the primitive method from the pay-as-you-go era.

Tsumagari, Mayumi, and Matsumoto Joji. 1972. *Budgeting in Japanese Companies.* Japan Productivity Center, pp. 222, 266.

CHAPTER 12

Globalization and the Future of Management Accounting

Since the late 1980s, the terms *global, international*, and *multinational* have appeared frequently in the Japanese press. It has influenced not only the Japanese economy but also Japanese accounting methods. As a result, management accountants have shown a greater interest in international accounting or management accounting in a globalized world than ever before. This reflects a remarkable increase in direct investment in overseas operations.

The purpose of this chapter is to discuss globalization in Japan and its impact on management accounting for foreign operations. We will first provide a short history of direct Japanese foreign investment abroad, describe the characteristics of overseas investment since 1985, and then discuss the management accounting issues that have arisen in Japan.

A Short History of Direct Foreign Investment Abroad

Japan began to invest abroad in 1951. Until the 1960s, however, most foreign investment was only intended to establish marketing channels to promote exports and to secure raw materials. Thus, the issue of globalization did not attract the attention of financial executives and academics, nor did it become a focus of management accounting in Japan until the 1970s.

The desire to develop production facilities abroad increased markedly during the 1970s, especially after the floating of exchange rates following Nixon's trip to China, and the 1973 oil crisis. These two events led to economic pressure on Japan, and attracted the attention of practitioners and academics in the area of international management accounting.

In the 1980s, direct investment abroad by Japanese companies grew rapidly due to the reduced competitiveness of domestic manufactures caused by the rise in the value of the yen. The yen rose from about ¥245 to the dollar to about ¥130 to the dollar after the meeting of the G5 countries in August 1985. In 1984, Japan's new foreign investments amounted to only $5,965 million. In 1990, only six years later, it had increased nearly eight times to $48,024 million. Though new investments dropped to $17,222 million in 1992, the mindset of Japanese managers toward large direct investments abroad has not changed.

The reason for this increase, simply stated, is that the price of foreign products had become significantly more competitive. As a result, in the late 1980s, the issues revolving around direct foreign investment became a major interest of Japanese business management. Japanese management accountants also became more interested in globalization.

In the 1990s, the yen continued to appreciate and in April of 1995 the exchange rate hit ¥80 per dollar. Rapid appreciation of the yen has made exporting difficult for Japanese companies and has increased the need for production facilities abroad, even in the strong automobile and electrical machinery industries. As a result, direct foreign investment continues at a high level, although it is not as large as it once was because of the bursting of the bubble economy and the resultant shortage of funds.

Since Japan began investing abroad again in 1951, there have been three main bursts of investment activity. The first was the boom period of 1972-1973, the second in 1978-1980, and the third from 1985-1989. Direct foreign investment since 1985 differs in several respects from that in the first and second boom periods. For many of the companies making investments in the third period, the investments were intended to be part of the core of the company rather than appendeges, which makes these investments the first step towards true globalization.

High Technology Products

In the past, the products Japanese companies produced and sold in overseas operations were not considered very important. These were products whose value-added levels were too low to produce price-competitively in Japan. However, the overseas investments which have grown most rapidly since 1985 have been in leading industries such as electrical machinery, transport machinery, and general machinery (MITI 1990).

It is predicted that the trend toward development of foreign operations by high technology companies will accelerate. The following are only a few examples (Sakamoto 1991). Matsushita Communications Industrial decided to transfer all exports from domestic production to overseas production. Sony and Canon increased the proportions of their production abroad from 40 to 60 percent and 40 to 50 percent respectively in 1992. Suzuki will reduce its exports from Japan to zero in the future. In January 1994, Nissan announced a planned increase in the proportion of its overseas production from the present one-third to two-thirds.

Small and Medium-Sized Enterprises

In recent years, the trend has changed from big businesses developing operations abroad to small and medium-sized enterprises developing foreign operations. For example, direct investment by small and medium-sized enterprises was only 33 percent (318 cases) of all direct investment in 1985 when the Plaza agreement was reached. It jumped to 60 percent of all foreign operations (1,625 cases) in 1988. Though the numbers have declined in later years (e.g., 1,401 cases in 1989 and 574 cases in 1992), the cumulative number of small and medium-sized enterprises with overseas operations has already exceeded 9,000.

Multinational Corporations

As Japanese overseas businesses slowly move towards becoming multinational corporations, they are reorganizing their management systems. For example, Matsushita Electric Industrial divided its worldwide operations into three distinct regions and each was given control over local operations in 1989. Sony, Yamaha, and Nissan have already developed four-unit systems for worldwide management.

In a four-unit worldwide system, the world market is divided into four blocks: a company with headquarters in the United States, a company with headquarters in Europe, a company with headquarters in Oceania, and the corporate headquarters in Japan managing local operations. The purpose of this reorganization is to shift finance, personnel, and management operations overseas, making the heart of operations local, and thereby gradually lessening the nationality of the company. The globalization of company management is expected to accelerate significantly as a result of the development of the four-unit worldwide system. For example, Mazda's sportscar, the MX-5 Miata, was designed in California, financed in New York and Tokyo, had the prototype produced in England, and is assembled in Michigan. Operations "transcend the bounds of the nationality" in this kind of multinational (Reich 1991).

Worldwide Scale

In the past, the majority of direct investment abroad was in Asian countries (Ministry of Finance 1990). For example, in 1977

Asia received the most direct investment, North America was next, followed by Europe. In Asia, this direct investment was concentrated in South Korea, Taiwan, and other NIE (Newly Industrializing Economies) countries.

Since 1985, investment has increased rapidly in North America and Europe. The amount of investment in the EC, and Eastern Europe in particular, has been tremendous. For example, in 1992 North America had the most direct investment followed by the EC and Asia. In addition, there has been a shift in direct investment in Asia from the NIE countries which have been growing rapidly and have had noticeable increases in wages compared to the ASEAN countries such as Thailand, the Philippines, Malaysia, and Indonesia. Investment also continues to grow in Oceania and in Central and South America. Though direct investment in Asia has increased again very recently, Japan's investment abroad continues to develop on a worldwide scale (JETRO 1994).

M&A Type Investments

Mergers and acquisitions (M&A) have been one of the typical strategies by which American and European firms have made inroads in foreign countries. In contrast, the typical approach by Japanese companies attempting to establish themselves abroad was to buy land overseas, and then to establish manufacturing plants and marketing bases there rather than acquiring companies through M&A.

Recently, in addition to establishing new factories abroad, the number of companies using M&A or engaging in joint ventures with local companies has increased. In fact, the number of M&As has increased three times in the five year period from 1985 to 1989. To use some familiar examples, in 1989 there was the acquisition of Columbia Pictures by Sony, the acquisition of the Manufacturers Hanover Trust subsidiary by Daiichi Kangyo Bank, the acquisition of Cook Cablevision by Sumitomo Shoji, and the acquisition of the Rockefeller group by Mitsubishi (Economic Planning Agency 1990). Although M&A by Japanese companies decreased remarkably in 1992 and 1993 because of poor economic conditions and the resultant shortage of funds, the impetus towards M&A by Japanese companies should be regarded as a continuing trend.

Resolving International Trade Frictions

Since 1985 there has been a major transition from exporting towards direct investment. The major direction in manufacturing has been to make direct investments that will reduce exports and increase employment abroad in order to resolve trade frictions. In particular, Japan's massive trade surplus was considered to be a problem by its trading partners in the 1990s. As a result of restrictions on imports, and the voluntary Japanese restrictions on exports, broad based export drives such as those conducted in the past are not likely to occur again.

As can be seen in the above discussion, Japan's direct investment overseas has grown rapidly and experienced changes. Though it decreased significantly for a few years after 1991 because of the bursting of the bubble economy and the resultant shortage of funds, it showed gradual recovery by 1994.

Management Accounting Issues in Globalized Enterprises

Accounting for companies operating abroad has been discussed either as issues of international accounting, accounting for globalization, or accounting for multinational enterprises, in Japan. In the past, the main theme was financial accounting. For example, although there is the Japan International Accounting Association, it is actually an International "financial" Accounting Association. Our interest in this book is different. We want to discuss issues of management accounting in globalized enterprises.

Management Accounting in Japan's Globalized Enterprises

Recently, management accounting in globalized enterprises has been called "Global Management Accounting" (MAFNEG

Research Association 1991a; 1991b). It tries to distinguish issues in global management accounting from the existing international financial accounting. Thus, for a moment we will discuss the words *global*, *multinational*, and *international* as they apply to Japanese companies.

The first question might be, "At what stage are Japanese companies presently, in terms of globalization strategy?" Iwabuchi (1993) tried a cluster analysis on companies listed on the Tokyo Stock Exchange using the Bartlett and Ghoshal (1989) framework:

International companies have a centralized core and the rest decentralized.

Transnational companies have specialized business units and a strong mutual inter-reliance in the company though it is decentralized.

Global enterprises are highly centralized and their overseas subsidiaries act as a pipeline in locally executing the strategy of the parent company.

Multinational companies are those decentralized companies in which the overseas subsidiaries have a great deal of autonomy.

According to the survey results, international and transnational firms are the most prevalent in Japan (37), followed by global enterprises (34). There are few multinational companies—only 13 in Japan. These are the companies that are thought to be able to respond quickly to special local demands. Judging from the above survey and the author's experience, we will conclude that a high percentage of Japanese companies can be classified as global enterprises.

Let us take a look at the relationship between the stages of international development and the issues for research in management accounting. Miyamoto (1983, 1989), in discussing the differences between international management accounting and multinational management accounting, described international management accounting as concerned with the issues of international transfer prices. Multinational management accounting, on the other hand, should deal with organizational structures, performance evaluation, the conversion of foreign currencies, price fluctuations, and risk in exchange rates. There have been various arguments advanced in Japan about these subjects, but at the moment the two factors which are transforming Japanese companies are globalization and high level computerization. In other words, the term "globalization" is used all-inclusively. Thus, in this book we will use the word globalization in this all-inclusive sense.

The Issues of Global Management Accounting in Japan

It is important to understand the current issues in global management accounting in Japan. The following is a listing of them in order of importance.

The investment justification of overseas operations. Three events trigger a need for investment justification: establishing a factory abroad, expanding or closing a factory, and M&A.

Performance evaluation in overseas operations. In such performance evaluation, there are two major issues: First, is the method of performance evaluation the same as is practiced in Japanese companies? Second, how should the evaluation of management and of the overseas operation itself be performed?

Overseas transfer pricing. Associated with this issue is the need to discuss the relationship between the domestic internal transfer value and international taxes.

Auditing affiliated companies abroad by the parent company. The way overseas auditing must be done is different from domestic auditing because of cultural differences.

Japanese management accounting of overseas operations. Transferring Japanese management accounting to overseas operations has become one of the urgent issues in operating successful overseas operations. For example, transfer of target costing to American and European companies has been going on for several years.

The relationship between the overseas company's accounting system and the operating system of the parent company. There will be cultural and institutional differences in accounting systems in different countries. For example, Japanese accounting organizations are typically organized by competitive-team orientation, not by command-and-control orientation (Sakurai 1994).

Avoidance of risk in the exchange rate, and country risk.

Internal reporting systems from overseas companies to the parent company. And conversely, those from the parent company to the overseas company.

In addition, the mid- and long-range business planning for global companies, the budgeting system for overseas operations, and the information systems of the global companies are factors in global management accounting. Finally, issues lying between financial accounting and taxation, segment reporting for foreign enterprise activities, price fluctuation accounting in relation to the exchange rates, and international taxation also deserve consideration.

Current Issues in Global Management Accounting

The performance of foreign investment is measured at two different times; the first is investment justification at the planning stage, and the second is performance evaluation after the investment. Investment justification at the planning stage precedes the investment decision.

How to Measure Efficiency of Foreign Investment

Efficiency of investment refers to return on investment (ROI). ROI in foreign investment is typically calculated by dividing the net profit, which is obtained by subtracting the capital costs from operating and other income by the average amount of investment during an accounting period. The amount of investment is the amount of invested capital in an overseas enterprise or the sum of the capital outlay and the amount borrowed. Both sales revenue and other sources of income such as dividends, royalties, and licenses should be included in the amount of return. Thus, the efficiency of investment is calculated according to the following formula.

$$\text{Efficiency of investment (\%)} = \frac{\text{(operating income + other income)} - \text{capital costs}}{\text{average amounts of (capital outlay + the amount borrowed) during an accounting period}} \times 100$$

The definition of ROI can vary depending on the circumstances. Most often it means cash returns to the parent company, but sometimes it means the change in value of the local firm. When the foreign investee has not itself become a global company, a simple evaluation comparing the investment and the return to the parent company may be sufficient. However, with increasing globalization, it is necessary not only to determine the amount of return to the parent company but the retained earnings of the local enterprise.

Methods for Justifying Investment in Foreign Operations

There are four major justification methods for foreign investment: ROI, payback, IRR (internal rate of return), and PV/NPV (present value methods). The net present value method is regarded as the best in the United States (Holland, 1986). Many cases have been cited of capital budgeting decisions using the present value method.

The MAFNEG study (1991) pointed out that in the United States it is common to eliminate political and social factors from investment justification of overseas operations. In contrast, it is customary for Japanese companies to take other social factors into account along with fluctuations of the exchange rate and the value of the yen. They give serious consideration to political, social, and other factors. Mr. Sugisaki, manager of the international taxation department in Toshiba, said that the most important considerations are how to procure necessary materials, plants, and equipment, not financial justification.

In the spring of 1994, the author visited three major Japanese companies (Nissan, Toshiba, and Sony) to conduct research on global management accounting. All the managers he met told him that they don't justify overseas investment based on financial considerations alone. In fact, financial justification seems to have a very low priority. Long-term perspectives dominate in these companies. When Japanese companies apply financial justification, they typically evaluate direct investments overseas on the basis of the payback period. Regardless of what method they use, investment justification of overseas operations may differ from that made domestically.

It is difficult to determine whether a project is feasible or not. At NEC Precision Equipment, they take such factors into consideration as the parent company's projected value, prospects of the world market, the operating plan, and the forecast in the budget (Shiba, 1989). Mr. Shiba at NEC Precision Equipment said, "Surprisingly, often such vague criteria as making an expected annual profit in the third year or having a cumulative profit by the fifth year are silently accepted without being clearly stated in typical Japanese companies." Since foreign investment by Japanese companies is not undertaken merely for short-term economic benfits, such tacit acceptance may be practiced by many companies.

In recent years many direct investments have been made abroad in attempts to avoid trade friction (Economic Planning Agency 1990). This has been especially true in the color television, video cassette recorder, and automobile industries. Thus, it is necessary to include other tangible and intangible effects besides profitability when making a decision on direct investment abroad. Among tangible effects are such outcomes as increases in sales volume, growth of market share, learning skills, technologies, flexibility to shift production, and improvement in quality. For intangible benefits there are such consequences as the easing of trade friction, leverage via host governments, contributions to the society of the host country, and improvement in the image of the company. Nissan, for example, developed operations in Great Britain because of the strong urging of Prime Minister Thatcher, allowing political judgments to take precedence over economic considerations. This certainly reflects the fact that investment justification for direct investments abroad cannot be gauged purely by economic results.

M&A is another way to establish a foreign presence when the know-how and talent in the target company are not available in the parent company. In justifying M&A, the payback period and other strategic factors are emphasized in Japan. For example, according to Mr. Miyabe, Executive Vice President of Mitsubishi Chemicals, the criterion for judging an acquisition is how long it will take to recover the amount invested in the purchased company. In the acquisition of Columbia Pictures by Sony, the acquisition of expertise in movie software was very important in terms of corporate strategy.

Performance Evaluation for Overseas Operations

There are two types of performance evaluation in foreign operations, just as in domestic operations. One is the evaluation of the local manager that is part of the system of management reporting. The other is the evaluation of the performance of the foreign enterprise itself.

When measuring the performance of a foreign operation, the currency units in the foreign operation usually differ from those used in the parent company. Thus, the financial performance of the company is always affected by fluctuations in the exchange rate. Since local management cannot control these fluctuations, it is necessary to separate the evaluation of the enterprise itself from the evaluation of the local manager. The most commonly used yardstick for performance evaluation in the United States is the comparison of budgeted and actual results (Kollaritsch 1984). Though a recent survey (Borkowski 1992) found that adherence to budget, while advocated in theory, is not a primary criterion in performance evaluation either domestically or internationally, a comparison between the budget and actual results was believed to be most appropriate to evaluate the performance of *managers* of foreign operations in the United States (Robbins and Stobaugh 1973). The most commonly used measure for evaluating foreign operations is ROI. There is empirical research (Borkowski 1992) that ROI is not the primary measure for multinational corporations. However, it is widely believed in the United States that ROI is the most appropriate means to rate the performance of a foreign operation itself (Choi and Mueller 1984).

Managers of Japanese companies, in contrast, place much less importance on the evaluation of the local manager. There is empirical data for supporting this inference. In Sato's survey (1991a), he asked the respondents to choose one of the following four reasons as the purpose of performance evaluation:

1. In order to know the state of the local company including non-accounting data
2. In order to evaluate the performance of the local manager
3. In order to confirm if local managers are producing appropriate levels of earnings
4. In order to check the appropriateness of the goals or strategy of local operations

Of the 129 companies that responded, the greatest number (50) replied that the most important reason was the level of earnings, followed by 44 companies choosing appropriateness of goals and 35 companies choosing the state of the local company. Not a single Japanese manager replied that the purpose of performance evaluation was to rate the performance of the local management.

When evaluating performance, which should be emphasized as a yardstick for accounting figures: ROI, ROS, profits, sales volume, comparison of budget and actual performance of ROI, a comparison of the budget and actual performance for profits, or a comparison of the budget and actual results for sales volume? According to Sato's survey (1991a), in the evaluation of the foreign enterprise itself, evaluation according to profit is most commonly used. This was followed by comparisons of the budget and actual performance, sales volume, and a comparison of the budget and actual results for sales volume. ROI was near the bottom. In other words, in Japan, ROI is not used much, in contrast to the United States and Europe. Conversely, what the foreign enterprise of a Japanese company emphasizes are profits and sales volume. Other than placing a slightly greater emphasis on evaluating the budget and actual results in the performance evaluation for managers, there is no significant difference between it and the performance evaluation of the enterprise itself.

Japanese managers believe that the performance of a foreign enterprise should not be judged simply on the basis of accounting numbers such as those mentioned above. The predominant view is that data not related to accounting, and contribution to society, should be included in evaluating the performance of a foreign enterprise. Data not related to accounting here refers to such data as quality, market share, company image, the labor relations environment, and the level of the managerial and operating system. Contribution to society refers to such things as the relationship with the local community, technology transfer, the number of locally hired personnel, environmental issues, and the amount of taxes paid to the local community. Most Japanese managers believe that it will be more and more important to emphasize contribution to society, as was suggested in Chapter 1.

Case Studies of Performance Evaluation in Japanese Companies

What sort of evaluations do Japanese managers practice in individual industries? We will introduce some examples based on publications (Business Research Institute 1991) and the author's experience of company visits.

Canon

At Canon, the company evaluates the records of foreign operations and the degree of contribution to the group based on the following three factors: (1) quantitative, (2) qualitative, and (3) an overall evaluation. In quantitative evaluations, it evaluates profitability, sales growth and potential for growth, productivity, efficiency, and contribution to society based on the following three tools: the income statement, the balance sheet, and the degree to which the budget has been achieved. Qualitatively the company evaluates the operation (not only operations but also operational efforts) and the degree of contribution to Canon as well as other qualitative factors. The company gives the qualitative performance a grade of superior (*yuu*), good (*ryou*), acceptable (*ka*), or unacceptable (*fuka*). For the overall evaluation, the company evaluates 15 criteria such as sales volume, rate of dividend, and inventory level, and then draws the result on a radar-chart.

Kikkoman

Kikkoman is the world's largest syoyu (soy sauce) brewer, enjoying a 30 percent domestic market share. It evaluates performance in both quantitative and qualitative terms. More specifically, the company evaluates performance on the basis of (1) return on invested capital, (2) the local growth rate , (3) the contribution to the basic policy of the corporation, and (4) an overall financial evaluation.

Kikkoman bases the evaluation of the dividend vis-a-vis the original purchase price of the securities in yen terms. The local growth rate of net worth is evaluated to avoid too much emphasis on the ROI. Policy is evaluated from the following four aspects: trademarks, technology, distribution, and new product development. Finally the overall evaluation is based on 10 indices such as the ROI and the ROS.

Oki Electric Industry

Oki is a leading manufacturer of communications equipment, with over a century of history in Japan. The performance of operations is evaluated from the viewpoint of earnings on capital invested by the parent company (including non-quantifiable factors), ROI, etc., as well as a comparison of the planned and actual performance by using such tools as the annual budget, and medium and long range planning. In addition, Oki evaluates the financial and nonfinancial performance of the local management on the basis of a comparison of projected and actual performance. For example, the company uses unit sales of a given machine and the achievement of yearly sales and budget projections.

Mitsui Bussan

Mitsui Bussan is a general trader with nearly 170 overseas offices. It evaluates the achievement level of operational plans and the ROI, but does not assign scores. It evaluates performance by placing emphasis on qualitative variables.

Why does Mitsui Bussan place emphasis on qualitative factors? It is because of the lessons learned from other Japanese companies in the same industry. That is, one of its competitors recently introduced overall evaluation systems based on three factors—ROI, trend analysis of the past three accounting periods, and an overall qualitative analysis. It was a complete failure. The sales managers opposed the system so strongly that it ceased functioning entirely.

As can be seen from the above, it is not always appropriate for Japanese companies to evaluate the performance of an overseas operation strictly on the basis of accounting numbers. It is essential that the overall performance be evaluated based on a number of different factors. However, the evaluation and measurement of qualitative factors is difficult. Thus, some companies have developed a clear formula for scoring qualitative factors that make a uniform evaluation by assigning weights. For example, Shiba

(1989) proposes a method to evaluate qualitative performance in three categories: good, acceptable, and unacceptable. Then, they make a chart to plot trends for making an overall evaluation and so predict the future.

Transfer Pricing in International Business

Until recently, transfer prices in domestic operations were considered to be an issue between the division and the parent company. The issue of transfer prices in international business resembles that of domestic transfer prices. However, one must consider several additional variables when determining transfer pricing in international business: difference in income tax between the two countries, custom duties, inflation, changes in currency exchange rates, export subsidies, expropriation, and level of competition.

Impact of transfer price on corporate income

Because tax policies differ from country to country, pricing of goods or services transferred to other countries is an important issue. Assume that a product is transferred to a tax haven country. If one follows a strategy of setting a price of the product and transferring it to a foreign country at a loss, it may result in quite a profitable business to a multinational company due to a reduction in taxes. Conversely, if transfer prices are set high, the amount of taxes paid in the country where the goods are received will be low and the amount of taxes paid in the country providing the goods will be high.

Some countries place restriction on the payment of profits and dividends. Thus, determination of the transfer price will significantly influence the profits of a company. In addition, one must also take customs duties and fluctuations in the exchange rates into account when considering international transfer prices.

Comparative study of international transfer price

According to the survey made by Tang (1979, 1981) on domestic transfer pricing and international transfer pricing, there is no significant difference between the United States and Japan with respect to domestic transfer prices and international transfer payment prices. A full cost-based policy was adopted by 42 percent of all respondents in the United States and 38 percent of those in Japan. For both countries the full-cost based method was the most common. The next most common policy was the use of market prices (37 percent in Japanese companies and 35 percent in the American companies). Only a few companies used negotiated prices in either Japan (22 percent) or the United States (14 percent). Though companies in Great Britain and Canada used the market price more than cost-based price as a basis for setting international transfer prices, there was no major difference in other respects from the practices of Japanese and American companies.

The study by Tang is valuable in that it is a comparative study. However, considering that the true globalization of Japanese companies began after 1985, more recent surveys are needed. According to the results of the survey on international transfer prices done by Sato in 1991, the cost-based method was most frequently used (58 percent) by Japanese companies. Specifically, the cost-plus-profit method was used by 47 percent (65 of the companies surveyed) and the cost method was used by 11 percent (15 companies). Market price (29 percent/39 companies) and market-price-minus basis (11 percent/15 companies) came next. We may conclude from his survey that the use of cost-based transfer prices has increased in recent years so that at present around 60 percent of the companies base their transfer prices on full cost in Japan.

How should international transfer prices be determined between the parent company and the foreign subsidiary? The research by Hoshower and Mandel (1986) disclosed that many U.S.-based multinational companies tend to have the international transfer price determined at the subsidiary level. What about in Japan? According to Sato (1991b), there was no company in Japan that allowed the foreign subsidiary to determine the international transfer price. The practice in Japan is significantly different from that of the United States. Only one-third of the Japanese companies responding (44 companies) had the international transfer prices determined by the parent company or the headquarters of the parent company. Most often it was determined by negotiation between the parent company and its subsidiary abroad (62 percent / 73 companies).

Auditing of Foreign Operations

Foreign business needs both external and internal auditing. Considering the vast resources controlled by multinational firms, the need for effective external auditing of these entities is clear (Hermanson 1993). Internal auditing of the foreign operations is conducted in three areas. First, checking to insure that there is no impropriety or error in order to preserve assets. Second, reliability checking to see that internal reporting is appropriate. Third, improvement in the managerial and operating systems. All of these fall within the area of internal auditing. While the first and second types are accounting audits, the third is an operational audit.

Auditing practices in the United States

In the United States, an internal audit of international operations is believed to be fundamentally the same as a domestic audit, although local accounting practices, foreign exchange, local operating practices, language, and customs need to be taken into account (Arpan and Radebaugh 1981). For example, it is argued that the following items must be checked in an internal audit of international operations (Evans, Taylor and Holzmannn 1985).

1. Be sure that training of the foreign personnel is appropriately given.
2. Be sure that regular meetings and seminars for foreign managers are held.

3. Be sure that the managers know the various regulations of the parent company.
4. Insure that the auditor possesses skills and familiarity with the local company, and
5. Be sure that corporate audit personnel make frequent visit to overseas subsidiaries.

We may conclude from the above references that the audit of overseas operations in the United States is an operational audit as well as an accounting audit.

Auditing practices in Japan

In internal audits of foreign operations, the operational audit is the primary method used in Japan. The term operational audit refers to the audit of operations which incorporates an accounting audit. Hida (1989) stated that the main themes of internal audits of foreign operations involve (1) a policy for reducing costs (taking advantage of appreciation of the yen against the dollar, and checking for measures for appropriate ordering, reduction in production costs, and increasing the turnover of capital), (2) the provision for and realization of a system for managing profits (a check on the effectiveness of finance activities), and (3) a reassessment of the subsidiary as a whole. Though these are chiefly audits of efficiency of operations which take accounting audits for granted, they may be considered a type of operation or management audit.

Nakao at Toshiba (1989) states from his experience that there are major differences between the internal audits conducted in the United States which focus primarily on checking for improprieties and mistakes and those conducted at Toshiba which place emphasis on operations. According to Mr. Nakao, the aim in auditing Toshiba operations abroad is to confirm the basic operating policy, insure reliability of the management system, ascertain the state of implementation of important operations, and emphasize communication.

Shiba at Nippon Electric Precision Instruments (1989) states that "What is needed most in Japan is to check the effectiveness of the establishment and operation of the administrative and management system of foreign operations and to check overall efficiency so that advice may be given for improvement." Thus, he adds, "it is desirable for both an accounting audit and an operational audit to be done particularly for foreign operations that have been given management responsibilities." From these examples it is easy to comprehend how much emphasis is placed on operational and management audits in Japan.

As the need for audits of foreign operations increases, the organization of internal auditing continues to change. At one company, domestic audits have been separated into three different jurisdictions. The organization for audits in this company is shown in Figure 12-1.

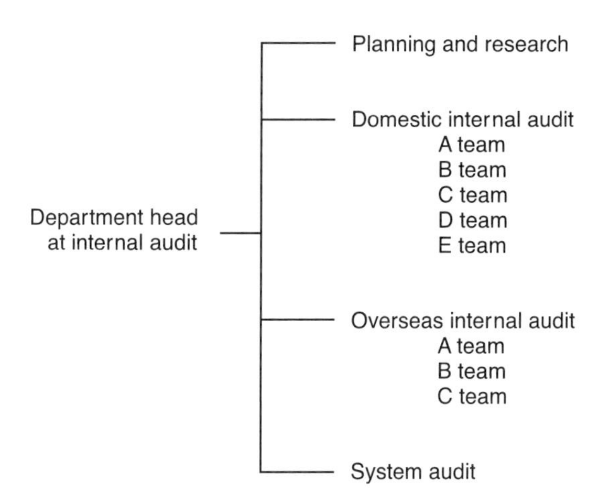

Figure 12-1. Organization for Internal Audit

In Figure 12-1, the testing group is not fundamentally different from an internal auditing group. Presently there are six people dealing with overseas operations. When necessary they receive the support of two staff from the section managing subsidiary companies overseas and three staff from the accounting department of the parent company.

Conclusion

Since the latter half of the 1980s, Japanese companies have initiated dramatic changes in their international strategies. They have changed from approaching the international market to increasing exports (an approach followed consistently by the government and people since the end of World War II) to approaching the development of overseas production facilities calling for globalization.

As a result, in terms of total assets abroad, Japan remains in third place behind the United States and Great Britain, while in terms of new investment, Japan was the country making the largest direct investments from 1989 to 1992. Investments are not focused in a single area but are being made on a global scale in such areas as North America, Europe, in the Newly Industrializing Economies and ASEAN nations, Oceania, and Latin America. These investments are not only in manufacturing but in finance, insurance, real estate, commerce, and in other non-manufacturing fields. Not only big business but small and medium-sized enterprises have joined the wave of globalization, reflecting the many faceted and increasingly complex nature of this phenomenon. It seems that true globalization has arrived in Japan.

References and Further Readings

Arpan, Jeffrey, and S. Lee H. Radebaugh. 1981. *International Accounting and Multinational Enterprises.* Warren, Gorham & Lamont, p. 328.

Bartlett, C. A, and S. Ghoshal. 1989. *Managing Across Borders: the Transnational Solution.* Harvard Business School Press. (Translation by Hideki Yoshiwara of *Company Strategy in the Age of World Markets*, Nihon Keizai Shinbunsha Inc, 1990).

Borkowski, Susan C. 1992. "International versus Domestic Managerial Performance Evaluation: Some Evidence." *The International Journal of Accounting* (December) pp. 129-139.

Business Research Institute. 1989. *The Globalization of Japanese Companies.* Business Research Institute Inc. (November) p. 43.

Business Research Institute. 1991. *Strategies of Companies with Operations Abroad and its Management*, 8th edition. Business Research Institute Inc. (July) pp. 101-233.

Choi, Frederick D. S., and Gerhard G. Mueller. 1984. *International Accounting.* Prentice-Hall, p. 408.

Economic Planning Agency Survey Bureau. 1990. *The Direct Investment Abroad Which Is Changing Japan and the World.* Ministry of Finance Printing Office, p. 27.

Evans, Thomas G., Martin E. Taylor, and Oscar Holzman. 1985. *International Accounting and Reporting.* Macmillan Publishing Company, pp. 322-323.

Hanada, Mitsuyo. 1991. "The Challenge to Control by Local Headquarters Presented by Divisions That Wish to be Independent." *Diamond Harvard Business* (Aug/Sept) pp. 69-78.

Hermanson, Dana R. 1993. "Multinational External Audit Planning." *The International Journal of Accounting* (June) p. 206.

Hida, Nobuo. 1989. "Global Strategy and Management Reviews." *QM* (QM Research Center, Chuo Coopers and Lybrand Accounting Firm) no. 12 (May) p. 74.

Holland, J. B. 1986. *International Financial Management.* Basil Blackwell, p. 105.

Hoshower, L. B., and L. A. Mandel. 1986. "Transfer Pricing Policies of Diversified Multinationals." *The International Journal of Accounting Education and Research* (fall) pp. 22-1.

Iwabuchi, Yoshihide. 1993. "Strategy Patterns of Companies Internationalizing and Performance Evaluation Systems." *Kaikei (Accounting)* Vol. 144, no. 19 (July) pp. 70-71.

JETRO (Japan External Trade Organization). 1994. *Overseas Direct Investment* (White Paper). p. 48.

Kollaritsch, F. P. 1984. "Managerial Accounting Problems of Multinational Corporations." In H. Peter Holzer, *International Accounting*. Harper & Row, Publishers, p. 190.

MAFNEG (Management Accounting for the Next Generation) Research Association. 1991a. "Management Accounting of Global Organizations." *Accounting* Vol. 43, no. 7 (July) pp. 128-134.

MAFNEG (Management Accounting for Next Generation) Research Association. 1991b. "The New Development in Management Accounting: In Search of a Global Management Accounting." *Industrial Accounting* Vol. 51, no. 2 pp. 100-110.

Miyamoto, Kanji. 1983. *Research in Basic Transfer Prices in International Management Accounting*. Chuo Keizai-sha, pp. 1-221.

Miyamoto, Kanji. 1989. *Management Accounting of Multinational Companies*. Chuo Keizai-sha, pp. 1-204.

Nakao, Norihiko. 1989. "Auditing Foreign Subsidiaries." *QM* (QM Research Center, Chuo Coopers and Lybrand Accounting Firm) no. 12 (May) pp. 79, 82.

Ministry of Finance. 1990. *Notifications of Direct Foreign and Direct Domestic Investments in 1989*. Ministry of Finance, pp. 2, 3.

MITI (Ministry of International Trade and Industry), Industrial Policy Bureau, International Enterprise Section. 1990. *Foreign Enterprise Activities of Japanese Companies*. Ministry of Finance Printing Bureau (March) p. 52.

Reich, Robert B. 1991. "Who is Them." *Harvard Business Review* (June-July). Translated by Yahara Tadahiko, "Global Companies Have Gone beyond the Nation State." *Diamond Harvard Business* (July) p. 4.

Robbins, Sidney, M., and Robert B. Stobaugh. 1973. "The Bent Measuring Stick in Foreign Subsidiaries." *Harvard Business Review* (September-October) pp. 80-88.

Sakamoto, Mitsuji. 1991. "Globalization and Regional Industry: Characteristics and Issues." In Japan Society of Business Administration. 1991. *Management Strategies for the 1990's.* Chikura-Shobo, p. 103.

Sakurai, Michiharu. 1994. "The Change of Accounting Organization and Its Practices." *Acccounting* Vol.46, no.10 (October) pp. 40-48.

Sato, Yaso. 1991a. "Management Accounting for Companies Incorporated Abroad." *Industrial Accounting* Vol. 51, no. 2, p. 62.

Sato, Yaso. 1991b. "Transfer Prices of Japanese Companies." *Accounting* Vol. 43, no. 10, pp. 68-69.

Shiba, Akira. 1989. "The Movement of Companies Abroad and Accounting." *Accounting (*August) pp. 131, 134, 136.

Tang, R. Y. W. 1979. *Transfer Pricing Practices in the United States and Japan.* Praeger, pp. 61-66.

Tang, R. Y. W. 1981. *Multinational Transfer Pricing: Canadian and British Perspective.* Butterworth & Co. Ltd., pp. 79-80, 127-129.

English/Japanese Glossary of Journals, Newspapers and Organizations

The following lists are original names of Japanese journals, newspapers and organizations and their translation as used in this book.

JAPANESE JOURNALS

Japanese Name	English Translation	Publisher
Ajiya Keiei Ronsyu	Management Review of Asia University	Asia University
Business Review	Business Review	Hitotsubashi Univeristy (Institute of Business Research)
Diamond Harvard Business	Diamond Harvard Business	Diamond Inc.
Genkakeisan-kenkyu	The Journal of Cost Accounting Research	Japan Cost Accounting Association

JICPA Journal	JICPA Journal	The Japanese Institute of CPA
Kaikei	Kaikei(Accounting)	The Moriyama Book Store
Kaikeijin Kousu	Accountant's Course	Chuo Keizai-sya
Kanrikaikeigaku	The Journal of Management Accounting, Japan	The Japanese Association of Management Accounting
Keieijitumu	Journal of Business Practice	Enterprise Management Association IMA Japan Chapter (or former NAA Tokyo Chapter)
Kigyoukaikei Management 21	Accounting Management 21	Chuokeizai-sha Japan Management Association
Matsuyama Daigaku Ronshu	Matsuyama University Review	Matsuyama University
Nikkei Jyouho Strategy	Nikkei Information Strategy	Nihon Keizai Shimbunsha Inc.
Sangyoukeiri	Industrial Accounting	The Industrial Management and Accounting Institute
Senshu Keiei-gaku Ronshu	Business Review of Senshu University	Senshu University
Senshu Keiei Kenkyujyo Shoho	Bulletin of the Institute of Business Administration	Senshu University
Senshu Syaken Nenpo	The Annual Bulletin of Social Science	Senshu University
Waseda Syougaku	The Waseda Commercial Review	Waseda University

JAPANESE NEWSPAPERS

Japanese Name	English Translation	Publisher
Nihon Keizai Shimbun(Nikkei)	The Nikkei	Nihon Keizai Shimbunsha Inc.

Nikkankougyou Shimbun	The Business and Technology Daily News	The Nikkan Kogyo Newspaper Company
Nikkeikinyu	The Nikkei Financial Daily	Nihon Keizai Shimbunsha Inc.

JAPANESE ORGANIZATION

Japanese Name	English Translation
Gijyutu Syuppan	Techno Publishing Co. Ltd.
Jouhou Sarvice Sangyou Kyoukai	JISA(Japan Information Service Industry Association)
Kigyoukeiei Kyoukai	IMA(NAA) Tokyo Chapter
Kigyou Kenkyukai	Business Research Institute Inc.
Kikaisinkou Kyoukai	Japan Society for the Promotion of Machine Industry
Nihon Jyouhousyori Kaihatsu Kyoukai	JIPDEC(Japan Information Processing Development Center)
Nihon Keiei Kyoukai	Japan Society of Business Administration
Nihon Kikaikougyou Rengoukai	The Japan Machinery Federation
Nihon Kikaku Kyoukai	Japanese Standards Association
Nihon Nouritu Kyoukai	Japan Management Association
Nihon Plant Mentenanse Kyoukai	Japan Institute of Plant Maintenance
Nihon Robot Kougyoukai	Japan Robot Association
Nikka Giren	Union of Japanese Scientists and Engineers
Sanno Daigaku	Sanno College Isehara (University) Sanno College Jiyugakuen (Junior C.)
Sangyoukeiri Kyoukai	The Industrial Management and Accounting Institute
Sangyoukouzou Singikai	Industrial Structure Council
Tsusansyou	MITI (Ministry of International Trade and Industry)

About the Author

Michiharu Sakurai

Michiharu Sakurai is professor of accounting in the School of Business Administration at Senshu University, Tokyo. He received his doctorate in Accounting from Waseda University and was a Fulbright Scholar at Harvard University from September 1989 to March 1990. He has served as a visiting faculty member to several U.S. universities including Stanford University from October 1992 to February 1993.

Professor Sakurai has published many books in Japanese including *Overhead Management* (1995), *Accounting for Software* (1993), *The Change of Business Environment and Management Accounting* (1992), *Management Accounting* (1990), *Cost Accounting for Software* (1987), *Cost Accounting* (1983), *A Study of Management Accounting Standards in the U.S.* (1982), and *Managerial Cost Accounting* (1979). He is coauthor of the English book, *Japanese Management Accounting: A World Class Approach to Profit Management*, with Professor Yasuhiro Monden, and has published numerous articles in both English and Japanese.

Professor Sakurai was voted the most creative and active accounting professor in Japan in a 1993 survey by Kawaijyuku. He is a recipient of the Ohta Award (Japan Accounting Association Outstanding Book Award) for 1995, the Keieikagaku Award (Japanese Management Science Literature Award) for 1992, the Gakujitsu Award (Japanese Public Accounting Academic Award) for 1982, and the Gakkai Award (Japan Accounting Association Outstanding Manuscript Award) for 1979.

He currently serves on a number of Japanese government and industrial committees related to the cost accounting and management accounting of high-technology, software, communication, and service industries.

Index

Books from Productivity Press

Cost Reduction Systems

Target Costing and Kaizen Costing

Yasuhiro Monden

Yasuhiro Monden provides a solid framework for implementing two powerful cost reduction systems that have revolutionized Japanese manufacturing management: target costing and kaizen costing. Target costing is a cross-functional system used during the development and design stage for new products. Kaizen costing focuses on cost reduction activities for existing products throughout their life cycles, drawing on approaches such as value analysis. Used together, target costing and kaizen costing form a complete cost reduction system that can be applied from the product's conception to the end of its life cycle. These methods are applicable to both discrete manufacturing and process industries.
ISBN 1-56327-068-4 / 400 pages / $50.00 / Order CRS-B259

Cost Management in the New Manufacturing Age

Innovations in the Japanese Automotive Industry

Yasuhiro Monden

Up to now, no single book has explained the new cost management techniques being implemented in one of the most advanced manufacturing industries in the world. Yasuhiro Monden has taught the principles of JIT in the U.S. and now brings us firsthand insights into the future of cost management based on direct surveys, interviews, and in-depth case studies available nowhere else.
ISBN 0-915299-90-9 / 198 pages / $45.00 / Order COSTMG-B259

Implementing a Lean Management System

Thomas L. Jackson with Constance E. Dyer

Does your company think and act ahead of technological change, ahead of the customer, and ahead of the competition? Thinking strategically requires a company to face these questions with a clear future image of itself. *Implementing a Lean Management System* lays out a comprehensive management system for aligning the firm's vision of the future with market realities. Based on Hoshin management, the Japanese strategic planning method used by top managers for driving TQM throughout an organization, Lean management is about deploying vision, strategy, and policy to all levels of daily activity. It is an eminently practical methodology emerging out of the implementation of continuous improvement methods and employee involvement. The key tools of this book build on the knowledge of the worker, multi-skilling, and an understanding of the role and responsibilities of the new lean manufacturer.
ISBN 1-56327-085-4 / 150 pages / $65.00 / Order ILMS-B259

Corporate Diagnosis

Meeting Global Standards for Excellence

Thomas L. Jackson with Constance E. Dyer

All too often, strategic planning neglects an essential first step—and final step—diagnosis of the organization's current state. What's required is a systematic review of the critical factors in organizational learning and growth, factors that require monitoring, measurement, and management to ensure that your company competes successfully. This executive workbook provides a step-by-step method for diagnosing an organization's strategic health and measuring its overall competitiveness against world class standards. With checklists, charts, and detailed explanations, *Corporate Diagnosis* is a practical instruction manual. The pillars of Jackson's diagnostic system are strategy, structure, and capability. Detailed diagnostic questions in each area are provided as guidelines for developing your own self-assessment survey.
ISBN 1-56327-086-2 / 100 pages / $65.00 / Order CDIAG-B259

New Performance Measures

Brian H. Maskell

Traditional performance measurements are not only ineffective for today's world class organizations, they can actually be harmful—they measure the wrong things. World class companies need measurements that can help them in their quest for improvement. You have to start measuring what your customers really care about such as customer service, quality, and flexibility. Implementing new continuous improvement programs while still using traditional performance measurements will only set you back and give you a lot of useless data. In *New Performance Measures*, you'll learn how to start measuring the things you truly need to know.

ISBN 1-56327-063-3 / 58 pages / $15.95 / Order MS4-B259

Performance Measurement for World Class Manufacturing

A Model for American Companies

Brian H. Maskell

If your company is adopting world class manufacturing techniques, you'll need new methods of performance measurement to control production variables. In practical terms, this book describes the new methods of performance measurement and how they are used in a changing environment. For manufacturing managers as well as cost accountants, it provides a theoretical foundation of these innovative methods supported by extensive practical examples. The book specifically addresses performance measures for delivery, process time, production flexibility, quality, and finance.

ISBN 0-915299-99-2 / 448 pages / $55.00 / Order PERFM-B259

REVISED!

20 Keys to Workplace Improvement

Iwao Kobayashi

The 20 Keys system does more than just bring together twenty of the world's top manufacturing improvement approaches—it integrates these individual methods into a closely interrelated system for revolutionizing every aspect of your manufacturing organization. This revised edition of Kobayashi's best-seller amplifies the synergistic power of raising the levels of all these critical areas simultaneously. The new edition presents upgraded criteria for the five-level scoring system in most of the 20 Keys, supporting your progress toward becoming not only best in your industry but best in the world. New material and an updated layout throughout assist managers in implementing this comprehensive approach. In addition, valuable case studies describe how Morioka Seiko (Japan) advanced in Key 18 (use of microprocessors) and how Windfall Products (Pennsylvania) adapted the 20 Keys to its situation with good results.
ISBN 1-56327-109-5/ 312 pages / $50.00 / Order 20KREV-B259

Handbook for Productivity Measurement and Improvement

William F. Christopher and Carl G. Thor, eds.

An unparalleled resource! In over 100 chapters, nearly 80 front-runners in the quality movement reveal the evolving theory and specific practices of world class organizations. Spanning a wide variety of industries and business sectors, they discuss quality and productivity in manufacturing, service industries, profit centers, administration, nonprofit and government institutions, health care and education. Contributors include Robert C. Camp, Peter F. Drucker, Jay W. Forrester, Joseph M. Juran, Robert S. Kaplan, John W. Kendrick, Yasuhiro Monden, and Lester C. Thurow. Comprehensive in scope and organized for easy reference, this compendium belongs in every company and academic institution concerned with business and industrial viability.
ISBN 1-56327-007-2 / 1344 pages / $90.00 / Order HPM-B259

Hoshin Kanri

Policy Deployment for Successful TQM

Yoji Akao (ed.)

Hoshin kanri, the Japanese term for policy deployment, is an approach to strategic planning and quality improvement that has become a pillar of Total Quality Management (TQM) for a growing number of U.S. firms. This book compiles examples of policy deployment that demonstrates how company vision is converted into individual responsibility. It includes practical guidelines, 150 charts and diagrams, and five case studies that illustrate the procedures of hoshin kanri. The six steps to advanced process planning are reviewed and include a five-year vision, one-year plan, deployment to departments, execution, monthly audit, and annual audit.
ISBN 0-915299-57-7 / 241 pages / $65.00 / Order HOSHIN-B259

Learning Organizations

Developing Cultures for Tomorrow's Workplace

Sarita Chawla and John Renesch, Editors

The ability to learn faster than your competition may be the only sustainable competitive advantage! A learning organization is one where people continually expand their capacity to create results they truly desire, where new and expansive patterns of thinking are nurtured, where collective aspiration is set free, and where people are continually learning how to learn together. This compilation of 34 powerful essays, written by recognized experts worldwide, is rich in concept and theory as well as application and example. An inspiring followup to Peter Senge's ground-breaking best-seller *The Fifth Discipline*, these essays are grouped in four sections that address all aspects of learning organizations: the guiding ideas behind systems thinking; the theories, methods, and processes for creating a learning organization; the infrastructure of the learning model; and arenas of practice.
ISBN 1-56327-110-9 / 575 pages / $35.00 / Order LEARN-B259

Measuring, Managing, and Maximizing Performance

Will Kaydos

You do not need to be an exceptionally skilled technician or inspirational leader to improve your company's quality and productivity. In non-technical, jargon-free, practical terms this book details the entire process of improving performance, from why and how the improvement process works to what must be done to begin and to sustain continuous improvement of performance. Special emphasis is given to the role that performance measurement plays in identifying problems and opportunities.
ISBN 0-915299-98-4 / 284 pages / $40.00 / Order MMMP-B259

A New American TQM

Four Practical Revolutions in Management
Shoji Shiba, Alan Graham, and David Walden

For TQM to succeed in America, you need to create an American-style "learning organization" with the full commitment and understanding of senior managers and executives. Written expressly for this audience, *A New American TQM* offers a comprehensive and detailed explanation of TQM and how to implement it, based on courses taught at MIT's Sloan School of Management and the Center for Quality Management, a consortium of American companies. Full of case studies and amply illustrated, the book examines major quality tools and how they are being used by the most progressive American companies today.
ISBN 1-56327-032-3 / 606 pages / $50.00 / Order NATQM-B259

TO ORDER: Write, phone, or fax Productivity Press, Dept. BK, P.O. Box 13390, Portland, OR 97213-0390, phone 1-800-394-6868, fax 1-800-394-6286. Send check or charge to your credit card (American Express, Visa, MasterCard accepted).

U.S. ORDERS: Add $5 shipping for first book, $2 each additional for UPS surface delivery. Add $5 for each AV program containing 1 or 2 tapes; add $12 for each AV program containing 3 or more tapes. We offer attractive quantity discounts for bulk purchases of individual titles; call for more information.

ORDER BY E-MAIL: Order 24 hours a day from anywhere in the world. Use either address:
To order: *service@ppress.com*
To view the online catalog and/or to order: *http://www.ppress.com/*

QUANTITY DISCOUNTS: For information on quantity discounts, please contact our sales department.

INTERNATIONAL ORDERS: Write, phone, or fax for quote and indicate shipping method desired. For international callers, telephone number is 503-235-0600 and fax number is 503-235-0909. Prepayment in U.S. dollars must accompany your order (checks must be drawn on U.S. banks). When quote is returned with payment, your order will be shipped promptly by the method requested.

NOTE: Prices are in U.S. dollars and are subject to change without notice.